FOUNDATIONS

ESSENCE BOOKS ON BUILDING

General Editor: J. H. Cheetham, A.R.I.B.A.

Other titles in the Essence Books on Building Series

ROBERT FISHER: Walls
G. HALE: Floors
R. E. OWEN: Roofs
H. WERNER ROSENTHAL: Structure
S. SMITH: Brickwork

Frontispiece. Failure of a foundation of an otherwise strongly built framed building after an earthquake in Japan.

FOUNDATIONS

V. C. Launder, A.R.I.B.A.
Senior Lecturer in Construction
Department of Architecture
University of Bristol

M

First edition 1972
Reprinted 1975 (with corrections), 1979

Published by
THE MACMILLAN PRESS LTD
London and Basingstoke
Associated companies in Delhi Dublin
Hong Kong Johannesburg Lagos Melbourne
New York Singapore and Tokyo

ISBN 0 333 13559 8

Printed in Hong Kong

Preface

No book of this size can possibly provide all the information necessary to design the foundations for any type of building. This has not been the intention.

The author has been more concerned with explaining the principles, and illustrating the balance of forces when the weight of a building is superimposed on the widely-varying geological strata of the earth's surface.

An introduction is given to the study of rocks and soils and methods of testing them, with a guide to the interpretation of test results.

Various types of foundation, and their applications, are described. Only simple design methods which illustrate the principles are given, but references are given for the further study of more complex examples. The emphasis has been on giving a wide outline of the subject, and, to use an apt metaphor, in providing a firm foundation on which to build!

Forwarned is forearmed, and the reader should recognize when the foundation problem is not a straightforward one. In such cases, special care in testing the soil and seeking the right solution is necessary.

Good practice in building construction is suggested for work below ground, in order to ensure stability and the control of moisture. This is necessary so that the principles can be related to the complex operations carried out on building sites in sequence. With a clear understanding of why and how a satisfactory solution to a constructional problem has been evolved, the more capable designer is well equipped to solve new problems as they arise.

V.C.L.

Acknowledgement

I am indebted to a number of people who have provided information and help over a long period. It is impossible to list everyone. Some have provided information indirectly through their books and other publications, or lectures, and might even be surprised to see their names here.

I apologize in advance to anyone who recognizes a phrase, or illustration which they originated and which I have used inadvertently without acknowledgement. In the technical field, a pioneer's research soon becomes the standard method, and this is also true in the educational field. Everyone draws on the work of his predecessors, and so progress goes on. The important thing is to collect the experience of others, add one's own to it and pass it on by some form of record, in this case a book, so that the knowledge is not lost, and further progress can be made by those who study it.

All the authors and publications mentioned in the lists of references have been of great value to me in my research for this book and will provide further sources of information for readers studying the subject more deeply.

Dr. Smith of the Dept. of Engineering, Bristol University and members of the staff of Messrs. Clarke, Nicholls and Marcel's Bristol office have helped check my methods of design and calculation in the examples given, for which I am grateful.

Picture acknowledgements

Frontispiece	Robin Phillips, Bristol
Figs 14, 15, 26, 27	B.R.C. Engineering Co. Ltd
Fig. 19	Pilcon Engineering, and Costain Photographic Library
Figs 20, 25, 30	Engineering Laboratory Equipment Ltd
Figs 50, 55, 59, 61	Frankipile Ltd
Fig. 52	British Steel Piling Co. Ltd
Figs 53, 54	Concrete Ltd
Figs 56, 60, 62	Peter Lind & Co. Ltd
Fig. 58	Taylor Woodrow Construction Ltd
Fig. 71	Kinnear Moody (Concrete) Ltd
Fig. 99	*Architecture and Building* July 1959
Fig. 104	Reproduction of newspaper photograph
Fig. 107	Millars Wellpoint International Ltd

Contents

List of Symbols	x
Introduction	xiii
SECTION ONE: THE SITE INVESTIGATION	
1. The site survey	3
2. Rocks and soils	5
3. Sampling and site testing	21
4. Laboratory tests on soils	39
SECTION TWO: THE FOUNDATION	
5. Principles of foundations	53
6. Selection of foundation type	61
7. Piles and piling	71
8. Retaining walls	97
9. Some special problems in foundation design	119
10. Preparing the site for foundations	129
11. Materials for foundations	145
Appendix: Examples of the design of simple foundations	149
Bibliography	157
Index	158

List of symbols

(Symbols mostly conform with B.S. 1991, Part 4)

As	Area of steel
B	Breadth of foundation
b	Breadth of wall
c	cohesion of soil at a given depth
c_u	cohesion – from undrained triaxial test
c_{cu}	cohesion – from consolidated undrained triaxial test
c_d	cohesion – from drained triaxial test
c_w	wall adhesion
c_o	cohesion at zero normal load
$\bar{c}$	average cohesion of soil
D	depth of foot of foundation below ground level
d	effective depth
E	Young's modulus
e	eccentricity
F	factor of safety
f	force
ff	frictional force
g	gramme
H	freestanding height of wall
h_w	vertical distance from bottom of foundation to ground water level
I	moment of inertia
k	kilo (1000)
K_a	Coefficient of active earth pressure (non-cohesive soils)
K_A	Coefficient of active earth pressure (cohesive soils)
K_p	Coefficient of passive earth resistance
$\overline{K}_s$	Coefficient of skin friction for piles
la	lever arm of reinforced concrete section
L	Length of foundation
LI	Liquidity Index of cohesive soils
LL	Liquid Limit of cohesive soils
M	bending moment
MR	moment of resistance
m	metre
m/c	moisture content
mm	millimetre
N	number of blows in a standard penetration test

N	Newton (a standard force equal to 1 kg × 1 m/s^2 acceleration)
O	Sum of perimeters of reinforcing bars in tension in a section of reinforced concrete
P	applied load or pressure
PI	Plasticity Index of cohesive soils
PL	Plastic Limit of cohesive soils
P_a	total active soil pressure
P_{an}	total active soil pressure normal to wall
P_p	passive earth resistance
P_{cb}	permissible compressive stress in concrete
P_{ct}	permissible tensile stress in concrete
P_{st}	permissible tensile stress in reinforcement steel
Q	maximum shear force
q	shear stress per unit area
q_a	allowable bearing capacity per unit area
q_s	safe bearing capacity per unit area
q_f	ultimate bearing capacity per unit area (failure stress)
R	resultant force
R_v	vertical component of resultant force
R_h	horizontal component of resultant force
t	tonne (metric)
V	volume
W	weight, concentrated
w	distributed weight or load per unit length or area
Z	modulus of section
z	vertical distance measured behind a retaining wall from base to top of retained soil.

Greek alphabet symbols

β (beta)	Inclination of a slope
γ (gamma)	density, bulk
γ_d	density of dry soil
γ_m	density of moist soil
γ'	density of saturated soil (submerged below water table)
γ_w	density of water
Δ (delta)	settlement
δ (delta)	angle of wall friction
μ (mu)	micro
π (pi)	3.142
ρ (rho)	settlement
Σ (sigma)	the sum of
ϕ (phi)	angle of internal friction
Ω (omega)	load factor

Introduction

Foundation design in the past has always been very much a matter of inspired guesswork on a background of past experience. Ancient methods were pathetically crude. Winchester cathedral was built originally on a foundation of short oak piles and bundles of wattles in a bed of peat but has survived on this for 750 years due to the preservative qualities of the peat. On the other hand, the famous Campanile at Pisa began to lean almost as soon as it was built and now has a dangerous overhang of 4.2 m out of plumb at the top. Today there would be no problem in rectifying this, but as long as it stays up, Pisa does not want to lose its greatest attraction to tourists; and all the engineers who see it as a challenge are constantly frustrated!

There was little scientific interest in the study of foundations until the spread of rational thought processes in France in the late 18th century led to the development by Coulomb of his theory of earth pressures against retaining walls. However, not until 100 years later was there a fresh surge of interest. This was due mostly to the problems arising from the construction of large commercial buildings on difficult sites, as in Chicago, where three "firsts" are claimed. The Borden Block (1880) is held to be the first building with independent footings taken down to sound strata below through a considerable depth of soft clay. The Montauk building (1882), 10 storeys high, was the first to use a steel grillage foundation, and the Stock Exchange building (1894) became the first to use large diameter concrete piles (eight wells were sunk to limestone bedrock and filled with concrete). As the height of structures soared, the need for more accurate methods of assessment for building foundations grew. Not until 1925, with the publication of "Erdbaumechanik" by Terzaghi, the father of modern soil mechanics, were scientific principles and methods made available to designers for the analysis and testing of soils, to establish their bearing capacity, settlement, and probable manner of failure.

A major step forward in this country was the establishment in 1930 of the D.S.I.R. Laboratory for research into soil mechanics. Since then, many international scientists have entered the field and important conferences on soil mechanics have taken place: at Harvard 1936, Rotterdam 1948, Zurich and Lausanne 1953, London

1957, Paris 1961 and Toronto 1965. There is now an International Society of Soil Mechanics and Foundation Engineering. The work of the Road Research Laboratory at Harmondsworth, Middlesex, has also been important in this field.

Today, we have many methods of checking the subsoil which were not available a few years ago, but test results can still only give us a very rough guide to the underlying strata, its strength and its probable performance under load. We can reduce the factor of safety (really a factor of ignorance), and build up a wide range of confirmatory evidence from all the means available, but there must always be a risk of some undetected weakness. This is a warning not to place implicit trust in the careful laboratory test, or in expensive site loading tests, but to use their results in the light of experience as useful aids to decision making in foundation design.

From the foregoing it will be apparent that there are two sides to foundation design:

(1) Exploration of the site to decide:

(a) if it is suitable to build on
(b) what the load-bearing capacity of the soil is likely to be
(c) where the best practical place to build the foundations is to be found (if, of course, there is room for choice).

(2) The selection and design of the type of foundation, necessary to take the load from the superstructure of the building and distribute it sufficiently to avoid overloading the soil; or in some cases, to anchor the building down.

SECTION ONE

The site investigation

1. The site survey
2. Rocks and soils
3. Sampling and site testing
4. Laboratory tests on soils

Chapter 1—The site survey

Before going on to the site, it is first necessary to be quite clear as to:

(a) *The client's brief,* the area of floor space required and how it can be disposed about the site, whether in single-storey or multi-storey building.

(b) *The exact location and boundaries* of the site.

(c) *The objectives of the site visit.* To return to the office and find some important piece of information missing can waste considerable time and money.

It is also necessary to have ready to hand all the equipment needed to carry out the survey. Check lists are valuable for both equipment and objectives. Such lists can be designed so that the information obtained can be written in on site, and will thus serve as a permanent record.

The main objectives with regard to foundations are:

1. To make an exact survey of the dimensions of the site.
2. To make an accurate survey of site levels, noting any difficult features such as steep slopes, hollows or holes which may have to be bridged.
3. To note obvious surface characteristics, such as visible soil and rock outcrops, trees and their positions, other vegetation, existing buildings, old walls and cellars, wells.
4. To consider water flow on and around site. If, for example, it is well drained or marshy (vegetation is a guide), the presence of springs, ponds, streams, rivers, tidal estuaries or sea; any of which may lead to flooding.
5. To look for signs of "trouble" – that is, signs of ground movement, cracks in the ground or surrounding walls or buildings,[1] evidence of mining operations,[2] altered levels or filled-in ground.
6. To check the existence and position of underground services.
7. To check if the subsoil could be utilized for construction purposes (e.g. sand and gravel might be used for concrete work).
8. To note the degree of exposure to wind forces.
9. To note where the foundations of neighbouring buildings occur on site boundaries.

10. To decide on possible positions for the building, having regard to its practical use, appearance, and the most economic arrangement for foundations (e.g. the balance of cut and fill desirable).
11. To discover the geological formation underlying the site and its bearing capacity.[3] This is the most difficult assignment, except possibly for some comparatively simple domestic type structure on a well-established and reliable formation. In such a case, one or two trial holes may suffice. In most cases however, soil investigation requires a great deal of expert time, trouble and expense before a reasonably accurate assessment of the strength and disposition of the underlying strata can be ascertained.

Notes

[1] Photographs should be taken of existing cracks and other signs of damage in neighbouring buildings to avoid possible claims for damages later, when building operations are under way. "Tell-tales" (patches of cement mortar, sometimes with a piece of glass embedded in them across an existing crack) serve as a useful check against further movement. B.R.S. now have an improved device for measuring movement (B.R.S. Current Paper No. 2/71).

[2] For this country, many records of mining activities are held by the National Coal Board. Local geological museums, and the Geological Survey also have records for reference.

[3] Ordnance Survey maps to large scale (1 : 1250, 1 : 2500, 1 : 10,000) and Geological Survey maps (1 : 10,560 and 1 : 63,360) are essential references. Apart from the Drift maps and Solid Geology maps, there are some horizontal and vertical sections, and "memoirs" recording much useful information which cannot all be shown on the maps. For England and Wales these are obtainable from the Director, Geological Survey and Museum, Exhibition Road, South Kensington, London, S.W.7. Local sources of information, such as the local Building or District Surveyor should also be consulted. With careful questioning even an hour or so in the local pub can often produce invaluable information on site conditions!

Chapter 2—Rocks and soils

In order to use the knowledge available of its general characteristics it is essential to identify the rock or soil on which the foundation will be built. So that we all speak the same technical language, soils have been classified into groups. This standard classification differs from that used by the geologist because the requirements of the engineer are very different. The latter needs simple means of field identification and a guide to strength related to structural characteristics. These are given in Table 1 of C.P. 2001 *Site Investigations* (reproduced by permission of the British Standards Institution) (Table 1). There is also a simplified version in the Building Regulations 1972 Table to Regulation D7 (Table 2) but the C.P. table is the best reference.

One way of classifying the many mixed soils lying in the area between clays, silts, loams and sands is the Triangular Diagram produced by the U.S. Public Roads Administration and shown in Fig. 1. Particle size is a further classification (see Fig. 2).

There is yet another soil classification system used mostly for road and airfield work and developed by A. Casagrande. This gives letter symbols to the types of soil, with a suffix letter indicating the grading and content of coarse-grained soils and the plasticity of fine-grained soils. These are given in Table 3 to help the reader understand any engineer's report quoting this classification. For example, a Casagrande classification of GU implies a uniformly graded gravel with little or no sand.

The complete details are given in Appendix E of C.P. 2001 and provide information on some properties such as frost-resistance, shrinking or swelling properties, drainage characteristics, bulk density range and suitable tests which are not given in the Table 1 classification.

Standard definitions of rocks and soils are given in Appendix C of C.P. 2001. It should be fairly obvious that a general knowledge of geology is useful, indeed essential, to the designer of foundations.

The major divisions of classification are:

(a) Rock.
(b) Coarse-grained, non-cohesive soils.
(c) Fine-grained, cohesive soils.
(d) Organic, fibrous soils.

Notes on each of these types follow.

Table 1. General basis for field identification and classification of soils C.P. 2001 : 1957

	Size and nature of particles				Strength and structural characteristics			
	1		Principal soil types 2	Composite types 3	Strength 4		Structure 5	
	Types		Field identification		Term	Field test	Term	Field identification
Coarse grained, non-cohesive	Boulders Cobbles		Larger than 8 in. in diameter Mostly between 8 in. and 3 in.	Boulder gravels Hoggin*	Loose	Can be excavated with spade. 2 in. wooden peg can be easily driven.	Homo-geneous	Deposit consisting essentially of one type
	Gravels		Mostly between 3 in. and No. 7 B.S. sieve	Sandy gravels				
	Uniform	Sands	Composed of particles mostly between No. 7 and 200 B.S. sieves, and visible to the naked eye. Very little or no cohesion when dry.	Silty sands	Compact	Requires pick for excavation. 2 in. wooden peg hard to drive more than a few inches.		
				Micaceous sands			Stratified	Alternating layers of varying types.
	Graded		Sands may be classified as uniform or well graded according to the distribution of particle size. Uniform sands may be divided into coarse sands between Nos. 7 and 25 B.S. sieves, medium sands between	Lateritic Clayey sands	Slightly cemented	Visual examination. Pick removes soil in lumps which can be abraded with thumb.		

Fine grained, cohesive	Low Plas- ticity.	Silts	barely visible to the naked eye. Some plasticity and exhibits marked dilatancy.* Dries moderately quickly and can be dusted off the fingers. Dry lumps possess cohesion but can be powdered easily in the fingers.	Clayey silts Organic silts Micaceous silts	Firm	Can be moulded by strong pressure in the fingers.	Stratified	Alternating layers of varying types.
	Medium Plas- ticity	Clays	Dry lumps can be broken but not powdered. They also disintegrate under water.	Boulder clays Sandy clays	Very soft	Exudes between fingers when squeezed in fist.	Fissured	Breaks into polyhedral fragments along fissure planes.
					Soft	Easily moulded in fingers.	Intact	No fissures.
			Smooth touch and plastic, no dilatancy. Sticks to the fingers and dries slowly. Shrinks appreciably on drying, usually showing cracks.	Silty clays Marls*	Firm	Can be moulded by strong pressure in the fingers.	Homo- geneous	Deposits consisting essentially of one type.
	High Plas- ticity		Lean and fat clays show those properties to a moderate and high degree respectively.	Organic clays	Stiff	Cannot be moulded in fingers.	Stratified	Alternating layers of varying types. If layers are thin the soil may be described as laminated.
				Lateritic clays	Hard	Brittle or very tough	Weathered	Usually exhibits crumb or columnar structure.
Organic	Peats		Fibrous organic material, usually brown or black in colour.	Sandy, silty or clayey peats	Firm	Fibres compressed together.		
					Spongy	Very compressible and open structure.		

* For definitions see Appendix C of C.P. 2001.

NOTE. The principal soil types in the above table usually occur in nature as siliceous sands and silts and as alumino-siliceous clays, but varieties very different chemically and mineralogically also occur. These may give rise to peculiar mechanical and chemical characteristics which, from the engineering standpoint, may be of sufficient importance to require special consideration. The following are examples:

Lateritic weathering may give rise to deposits with unusually low silica contents, which are either gravels or clays; but intermediate grades are rare.

Volcanic ash may give rise to deposits of very variable composition which may come under any of the principal soil types.

Deposits of sand grade may be composed of calcareous material (e.g., shell sand, coral sand) or may contain considerable proportions of mica (where grain shape is important) or glauconite (where softness of individual grains is important).

Deposits of silt and clay grade may contain a large proportion of organic matter (organic silts, clays, or muds) and clays may be calcareous (marls).

Table 2. Ta
(Minimum wi

(1) Type of subsoil	(2) Condition of subsoil	(3) Field test applicable
I Rock	Not inferior to sandstone, lime-stone or firm chalk	Requires at least a pneumatic o other mechanically operated pi excavation
II Gravel . . . Sand	Compact	Requires pick for excavation. Wooden peg 50 mm square in cross-section hard to drive beyo 150 mm
III Clay . . . . . Sandy clay . .	Stiff	Cannot be moulded with the fingers and requires a pick or pneumatic or other mechanical operated spade for its removal
IV Clay . . . . . Sandy clay . .	Firm	Can be moulded by substantial pressure with the fingers and ca excavated with graft or spade
V Sand . . . . . Silty sand . . Clayey sand . .	Loose	Can be excavated with a spade. Wooden peg 50 mm square in cross-section can be easily driven
VI Silt . . . . . Clay . . . . . Sandy clay . . Silty clay . . .	Soft	Fairly easily moulded in the fingers and readily excavated
VII Silt . . . . . Clay . . . . . Sandy clay . . Silty clay	Very soft	Natural sample in winter condi exudes between fingers when squeezed in fist

Rock

Hardest rocks are the *igneous* rocks such as granite and basalt. These may have 2-3 times the safe bearing capacity of hard sedimentary rocks and 25-50 times that of clays or sands. With *sedimentary* rocks however, much depends on the angle of stratification and their cementing material, while some, such as chalk, can soften and weaken in wet conditions. Large cavities and swallow-holes (Fig. 3)

ion D7
ındations)

	(4) Minimum width in millimetres for total load in kilonewtons per lineal metre of loadbearing walling of not more than:				
/m	30 kN/m	40 kN/m	50 kN/m	60 kN/m	70 kN/m
	In each case equal to the width of wall				
	300	400	500	600	650
	300	400	500	600	650
	350	450	600	750	850
	600	Note: In relation to types V, VI and VII, foundations do not fall within the provisions of regulation D7 if the total load exceeds 30 kN/m.			
	650				
	850				

can also occur in sedimentary rocks and only careful investigation can detect them. Fissures may need filling with concrete or bridging over, although fissured rock is easier to split out by driving in steel wedges when excavation is required (see Chapter 10).

Generally, bedrock is an excellent base to build on, and often no other foundation is required. If it has to be levelled, excavated, or trenched for services, however, it can prove expensive.

Table 3. Casagrande soil classification

Prefix	*Suffix*
G Gravel	*Coarse-grained soils*
S Sand	W Well-graded with little or no fines
M Silt	C Well-graded with suitable clay binder
C Clay	U Uniformly graded with little or no fines
O Organic silts and clays	P Poorly graded with little or no fines
Pt Peat	F Poorly graded with appreciable fines or well-graded with excess fines
	Fine-grained soils
	L Low compressibility
	I Medium compressibility
	H High compressibility
	Fibrous soils – no sub-division

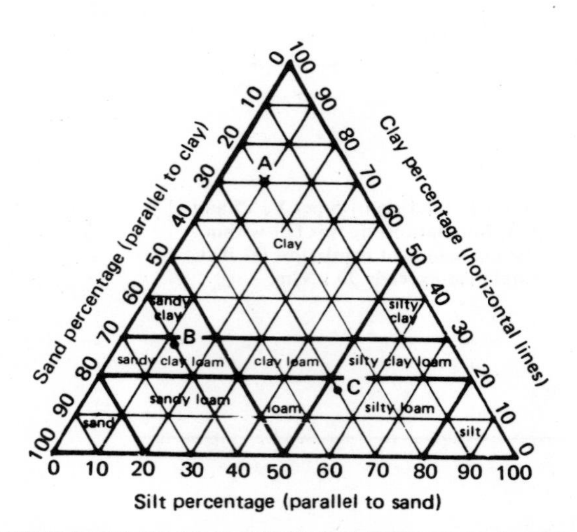

TYPE OF SOIL CAN BE PLACED BY ANALYSING % INGREDIENTS (PARTICLE SIZES)

E.G. SOIL A CONTAINS 70% CLAY 20% SAND 10% SILT AND IS CLASSIFIED AS A CLA
SOIL B CONTAINS 27% CLAY 60% SAND 13% SILT AND IS A SANDY CLAY LOAM
SOIL C CONTAINS 17% CLAY 30% SAND 53% SILT AND IS A SILTY LOAM

Fig. 1. Triangular diagram for soil classification (US Public Roads Administration).

Log scale mm	2	1	0	-1	-2	-3	-4	-5	-6
Particle size	millimetres 100	10	mm 1	micrometres 1000	100	10	µm 1	nanometres 1000 100	10 nm 1

Standard classification											
	200	20.0	6.0		0.6	0.2		0.02	0.006 mm		
	Cobbles	Gravel C	M	F	Sand C	M	F	Silt C	M	F	Clay
	60.0			2.0			0.06			0.002 mm	

Scientific description				
	Macroscopic		Microscopic	Sub-microscopic
	Very Coarse	Coarse	Fine	Very fine	Colloidal

SOILS WITH MORE THAN 10% RETAINED ON A 40mm B.S. 410 SIEVE ARE CONSIDERED AS AGGREGATES AND COVERED BY B.S. 812

Fig. 2 Classification of soils by particle size. (After Terzaghi and Peck.)

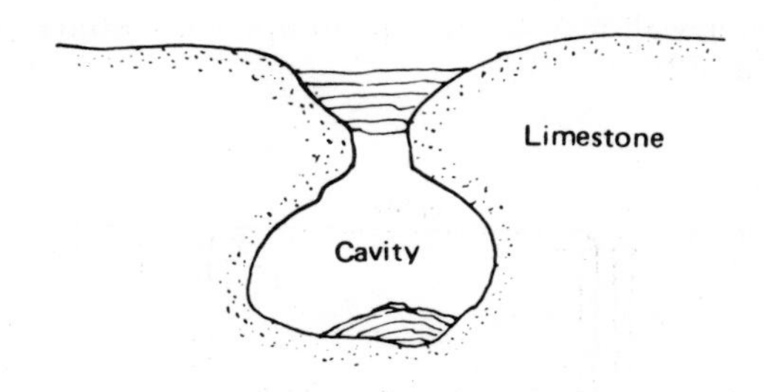

Fig. 3. Swallow-hole in limestone.

The comparative safe bearing capacities of rocks are given in British Standard C.P. 101 *Foundations and Substructures* as:

1. Igneous and gneissic rocks in sound condition	10,700 kN/m^2
2. Massively bedded limestones and hard sandstones	4 300 kN/m^2
3. Schists and slates	3 200 kN/m^2
4. Hard shales, mudstones and soft sandstones	2 200 kN/m^2
5. Clay shales	1 100 kN/m^2
6. Hard solid chalk	650 kN/m^2
7. Thinly bedded limestones and sandstones	To be assessed
8. Heavily shattered rocks and the softer chalks	after inspection

These figures are intended only as a guide and must be backed by other tests.

Rock pinning is sometimes necessary to prevent inclined strata from shearing and sliding under the increased load from new building work. The rock is drilled out to a calculated safe depth through the unsafe strata, after which reinforcement is inserted and concrete poured in. When set, this provides a "pin", resisting movement of the top strata. As many of these pins are provided as are necessary to stabilize the ground.

Rock bolts with wedges at the ends, or with expanding ends, are used to fix unstable rocks back to sounder rock, and sometimes long anchor bolts or steel bars are fixed in drilled holes in the rock and grouted in (Fig. 79 illustrates a similar technique).

A more recent technique is known as rock stitching. A line of holes is drilled into the unsafe strata, and steel cable is then looped into each hole in a continuous length and grouted in (Fig. 4) or bonded in with plastic resin.

A retaining wall may be used to stop inclined strata from sliding forward (Fig. 82).

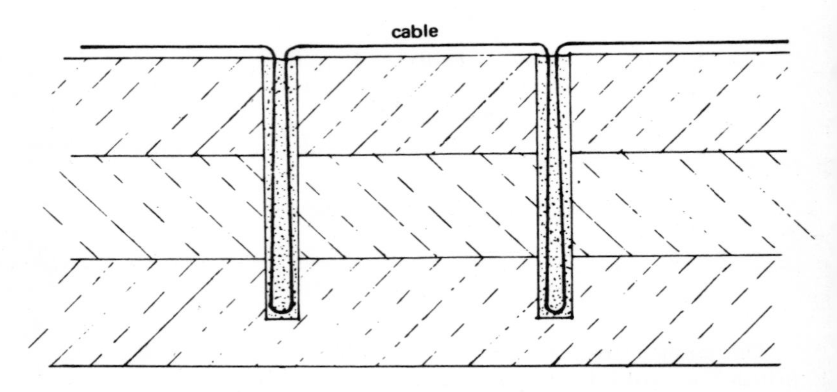

(From paper by K. G. Stagg, B.Sc., Ph.D., F.C.S., on stressed ground anchors)

Fig. 4. Rock-stitching technique.

Coarse-grained non-cohesive soils

Gravels and sands come under this heading. When loaded they can fail by shearing if unconfined, by the particles sliding over each other at an angle known as the angle of internal friction ϕ. This angle varies with the angularity or roundness of the particles. By analogy, a pile of children's playbricks will form a steep pile, but a pile of glass marbles will collapse flat. To take the analogy further, by confining the glass marbles in a strong box, it would be possible to stand safely on them, since they could not collapse if tightly packed.

Similarly, *confined* compacted sand or gravel can form excellent ground on which to build. The greater the load, the more the frictional strength.

Settlement will occur when such soils are first loaded, but there will be little change after this initial settlement, and no "recovery" if the load is removed. Sand can be vibrated to a more dense mass, which will reduce its surface level (similar to settlement) and increase its load-bearing capacity.

Gravel has the useful characteristic of being able to distribute uneven loading over a wide area. This fact is made use of in areas where mining subsidence is expected by the provision of a deep man-made gravel layer between foundation and the unstable ground.

Both sand and gravel are permeable (Fig. 5), so where the water table is high and there is lateral pressure on a basement from these soils, there will also be additional pressure from water saturating the ground, both laterally and vertically upwards underneath.

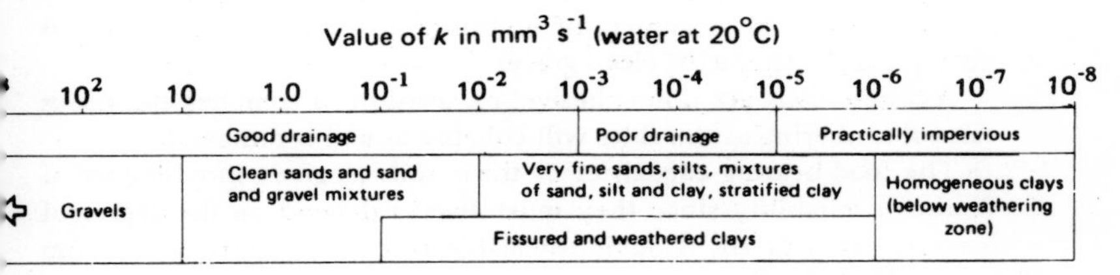

(Casagrande & Fadum)

Coefficient of Permeability k = rate of flow of water at a standard temperature per unit area of soil under unit hydraulic gradient and is expressed as a velocity in mms/sec.

Fig. 5. Permeability of soils.

Providing it is confined, sand will become more bulky (up to 25% greater volume at 6-8% m.c.) when damp and capable of resisting greater loads than when dry. Saturation can, however, halve bearing capacity, as so much water reduces friction by forming a film of water around each grain.

A saturated fine sand *unconfined* and subject to a hydraulic gradient (for instance, at the foot of an impermeable slope with water running down it) can develop into a quicksand (Fig. 6), as the upward pressure of water through the sand can cause it to "boil" or become incapable of sustaining anything but a very light load. Nevertheless, such a sand, if properly drained and consolidated, would sustain normal loading.

Pumping water out of sand to de-water a site which is saturated will not only cause settlement through consolidation but also settlement by the sucking out of sand with the water, unless filters are used to prevent this occurring. Existing structures bordering a

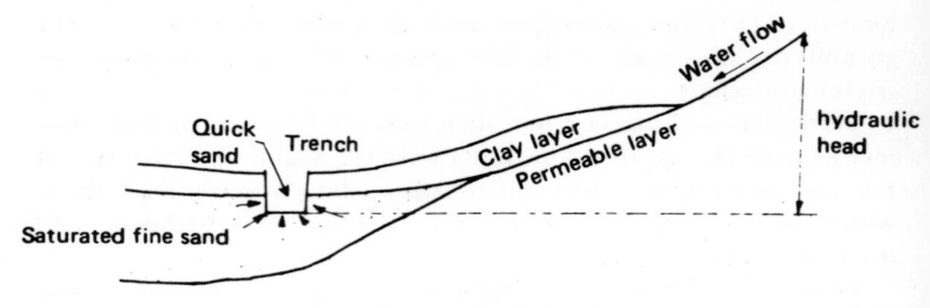

Fig. 6. Development of quicksand situation.

building site have been damaged in this way. It often requires very powerful pumps indeed to deal with the quantities of water which can flow through the permeable soil and fill an excavation rapidly.

On the other hand, where the water table is low, the coarser sands and gravel are self-draining, providing dry building sites and healthy conditions for the occupiers of buildings erected on them. There is no capillary action with clean gravel.

As these soils are non-cohesive, excavation of trenches etc. needs careful timbering as the sides will collapse as soil is removed.

The load-bearing capacities of these soils as given in tables are of doubtful reliability since they must always depend on the degree of wetness. This factor is obviously liable to change, quite apart from other variables, but as a rough guide to comparative strengths, the following safe bearing strengths taken from C.P. 101 may be of use:

	Dry	*Submerged*
1. Compact well-graded sands and gravel-sand mixtures	430-650 kN/m^2	220-320 kN/m^2
2. Loose well-graded sands and gravel-sand mixtures	220-430 kN/m^2	110-220 kN/m^2
3. Compact uniform sands	220-430 kN/m^2	110-220 kN/m^2
4. Loose uniform sands	110-220 kN/m^2	55-110 kN/m^2

"Dry" means at least a foundation-width in height above the water table. Foundation widths must not be less than 900 mm before the above figures can be applied.

Unless confined, loose dry uniform sands can be a dangerous foundation, as they can be eroded by winds. Thus, building on sand dunes is a risky business!

Saturated sand can contain half its volume of water. It will therefore be greatly affected by freezing conditions in winter, when it will heave upwards for a distance of 4.5% of the depth to which it is frozen. A situation in which this sort of problem can occur and build up to give a lot of trouble is under the floor of a cold store. In such locations thick thermal insulation is essential to prevent floor heave.

Fine-grained cohesive soils

These soils consist of very fine microscopic particles which become sticky when wet and hold together when dry. Some types become fissured by shrinkage as they dry out, and the surface can crumble when exposed to the normal weathering agencies. Clays and silts come under this heading. Such soils can absorb water by capillary action, so that clay under a building will still be damp. Seasonal changes in the water content of clay cause shrinkage and swelling, the latter exerting considerable pressure, sufficient to crack shallow foundations and the walls above (Fig. 7). A record of the annual

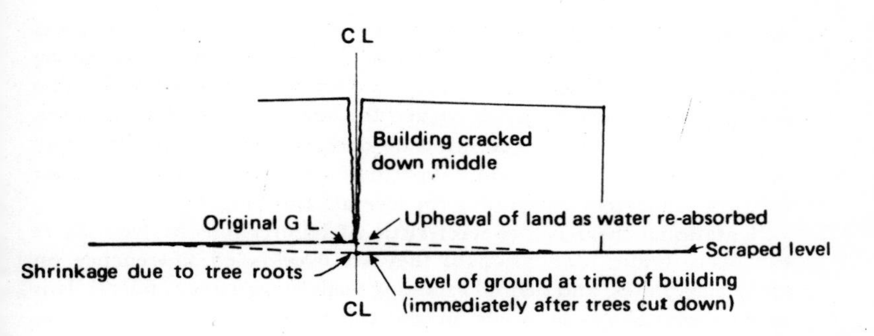

Fig. 7. Effect of swelling of clay soil after removal of trees and reabsorption of water. Building on site too soon can cause failure in this case.

vertical movement of foundations resting on London clay was made and resulted in:

13 mm with foundation at 900 mm depth
25 mm with foundation at 600 mm depth
38 mm with foundation at 300 mm depth

For this reason, foundations on clay should be placed at least 1 m below ground level.

Frost action can also cause upward pressures in clay and silty soils due to the formation of ice "lenses" which "grow" when fed by water, thus causing localized upheaval (Fig. 8).

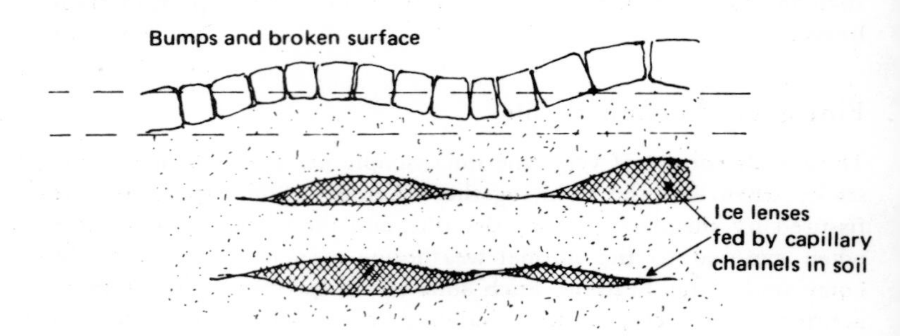

Fig. 8. Formation of ice lenses underground leading to frost heave of surface. (From paper by Prof. Everett, Bristol University, *New Scientist* January 1962.)

Providing that water is not squeezed out, clay will compress under load, and expand when the load is removed, the action being analogous to a hydraulic railway buffer with a slow leak. The re-expansion is very noticeable when the heavy weight of overburden is removed in a deep excavation. Squeezing out of water laterally under the weight of a building superimposed on the soil is the reason for slow settlement continuing for several years (Fig. 9).

Rotational shear is a characteristic of failures in cohesive soils. It can occur under foundations if they are overloaded, in trenches, on steep clay banks, and behind retaining walls (Figs 10, 11, 65(c)). It is particularly likely to occur in the last three cases when the clay develops deep fissures at the top of possible shear planes, thus reducing the area of cohesion which normally counterbalances the load, and tending to cause failure (see Fig. 11). Lubrication, by water seeping in, lowers the cohesion value of the soil and adds to the risk of failure. Cohesive soils cannot be quickly compacted by vibration methods. On the other hand, vibration or squeezing under load can cause liquid clay to flow, or the liquid to be squeezed out sideways, thus producing slow settlement over a long period. This effect can be dangerous when heavy vibrating machinery is located in a factory built upon a clay subsoil.

There are many types of clay, including one known as *boulder clay*; this contains quite large stones, dispersed within its strata. Permissible bearing pressures vary quite widely. Tests must be made to check the strength of a particular soil, but as a guide to probable safe bearing capacities the following will be of help:

1. Very stiff boulder clays and hard clays with a shaley structure — 430-650 kN/m^2
2. Stiff clays and sandy clays — 220-430 kN/m^2
3. Firm clays and sandy clays — 110-220 kN/m^2
4. Soft clays and silts — 55-110 kN/m^2
5. Very soft clays and silts — 0-55 kN/m^2

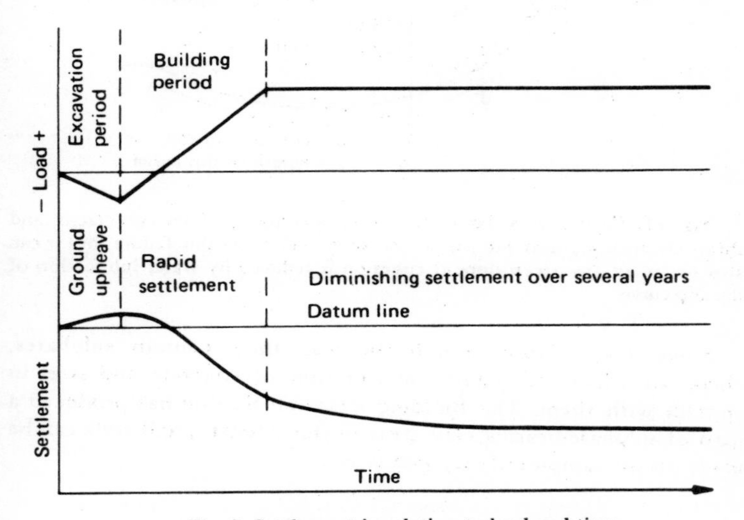

Fig. 9. Settlement in relation to load and time.

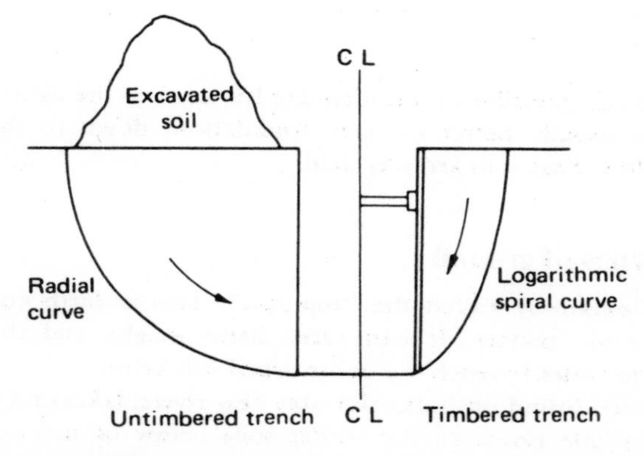

Fig. 10. Collapse of trenches due to rotational shear of clay soil.

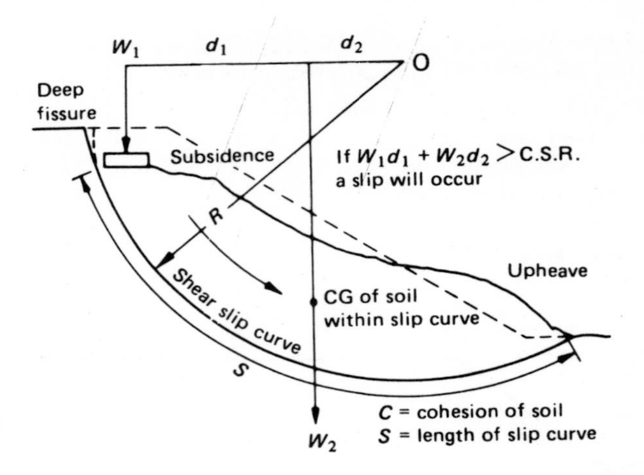

Fig. 11. Failure of a clay bank due to opening up of tension cracks, and shear. Overloading near the top of the slope will cause this failure, but it can also occur on any steep slope if cohesion is reduced by water lubrication of the slip curve.

Some clays cause trouble because they contain sulphates, chemicals which can cause deterioration of concrete and steel in contact with them. The Building Research Station has produced a map of sulphate-bearing clay areas in this country, and tests can be made on site samples of clay (see later).

Mixed non-cohesive and cohesive soils

These can nearly always be treated as cohesive, as it only requires a small percentage of clay (say 10%) in a soil to make it plastic.

Peat

This is so compressible and inadequate for any but the lightest loads, that it is usually better to carry foundations down to the firmer strata below. Peat soils are very acid.

Other types of ground

Loam Commonly called the "top spit". This is earth containing much organic matter. It can carry little weight and should be removed in order to reach the undisturbed soil below.
Mud Often found on industrial sites near rivers, lakes and sea. One can either pile down to the harder soils below or use a buoyant foundation (see later).

Made-up ground

Typical examples of made-up ground are lands reclaimed from sea or marshes, and disused refuse tips. Its bearing capacity is always suspect, and will depend on the fill materials, method of filling and degree of consolidation, not forgetting the strength of the underlying soil. Most refuse tips are very unsafe due to the voids formed by old cans and drums etc. which in time will collapse under load. The constituent materials may be combustible, and once alight, will prove difficult to extinguish. From the point of view of health such soils may also not be good for human occupation for a long period of time.

Where fill is necessary to build up a site, it can be made satisfactorily load-bearing by using good solid inorganic fill, natural or industrial waste material. This is put down in layers not exceeding 300-450 mm thick, consolidated by heavy rolling before the next layer is placed.

Apart from this type of fill, it is generally safer to transfer building loads direct down to the underlying strata, usually by piling (see later).

Chapter 3—Sampling and site testing

Soil investigation

Soil investigation may be carried out by:

(a) *The architect,* if a simple straightforward domestic size structure is to be built on known reliable soil.
(b) *A consultant engineer.*
(c) *A specialist site-investigation contractor* undertaking site boring, sampling, testing (in own laboratories).
(d) *A piling contractor.*
(e) *An independent private testing laboratory.*
(f) *University or Technical College test laboratories.*
(g) *A specialist engineer on the staff* (large organizations only).

The architect and consultant engineer may well only obtain samples from the site to send to a testing laboratory such as those run by (e) and (f) who, for their part, may not be prepared to carry out the actual site sampling. Generally speaking, this arrangement is not as satisfactory as (c) above, for when the whole job of sampling and testing is done by one organization, the overall picture can be seen, and evidence noticed on the site can be tied up with the results of laboratory tests. The responsibility is not split, and there is less chance of faulty sampling or of a mistake remaining undiscovered. If a piled foundation is obviously going to be required and there is to be no competitive tendering for the job, the alternative (d) may well be satisfactory. The last alternative (g) is common practice among county and city councils, very large architectural practices and the larger firms of building contractors.

The cost of a small soil investigation is likely to be 1-2% of the cost of a contract, with a minimum of about £250, if carried out by a specialist consultant. This outlay may, however, result in savings over the conservative foundation sizes given by rule-of-thumb design methods. It is certainly a wise insurance policy against trouble later.

Experienced men can often get useful information from the most primitive tests. They will judge the type of clay by the imprint of a heel or the penetration of the blunt end of a pencil. However, the usual site test for a small job is to dig one or more trial holes to the depth intended for the foundation base and then carry out a simple load test in each hole, or take away undisturbed samples for testing.

Undisturbed samples are essential for strength tests but obtaining them requires great care, as any sampling tube pushed into the soil is bound to alter the edges to some extent. A simple method is to cut a step in the bottom of the trial hole (which is usually about 600 mm × 1200 mm and at least 900-1200 mm deep) and then cut out a chunk of the step with a spade. An alternative method is to force a bottomless tin into the bottom of the hole, cut away all round, and slice the surplus soil away from the underside with a thin wire. A third method is to use a hand auger to excavate a hole to the required depth (3-6 m is possible with this tool in soft soils), and then replace the auger cutting end with a split tube sampler (Figs 15 and 16) or a "Shelby" thin-walled metal sampler tube which is forced into the ground and then withdrawn. These are cohesive soil techniques, as non-cohesive soils would break up.

Trial pits should be dug at least one every 900 m^2 but more may be thought advisable (e.g. one every 15 m in each direction). Holes should not be dug where the foundations will actually be placed, but near to important points, as backfill provides poor foundation conditions if the trial holes were to go too deep and require filling.

In order to check underlying strata for weaknesses or cavities, it is best to bore a deeper hole in the middle of the trial pits to a depth at least equal to the width of the building. Specimens of undisturbed soil can be taken for testing from any level. Bore hole sizes should be related to the type of soil e.g. 38 mm or 100 mm are usual sizes for clays, but 150 mm size may be required for gravels. It is not necessary to go down very far if bedrock is encountered, but some drilling into the rock (say 3 m) is desirable to establish that it is thick bedded and not, for example, merely an isolated stone in boulder clay. Rotary diamond drilling is necessary for the penetration of rock.

Wash boring (Fig. 16) is often used on larger projects, and can eliminate the large trial hole. It is carried out by driving into the ground lengths of 63 mm diameter pipe, screwed together, with a narrow diameter wash pipe down the middle. On the end of the wash pipe is fitted a chopping bit which breaks up the soil when the wash pipe is rotated, and lifted up and dropped. The wash water escapes up the annular space to the top, carrying with it the loosened soil which can be examined as it emerges into the water tub. This system obviously does not provide undisturbed samples, but the chopping bit can be withdrawn at intervals and samples taken by screwing on a "split spoon" and forcing it into the bottom of the hole. It is usually possible to make a fairly accurate vertical section through the strata penetrated, providing it is not rock.

Having obtained borehole sections at fairly regular spacings, a soil profile can be drawn, i.e. a section through the site showing the underlying strata, from which suitable depths of foundations can be decided (Figs 17 and 18).

Information from trial holes

The information required from trial holes and boreholes is as follows:

1. Thickness of layer of top spit.
2. Strata pattern, thickness and depth.
3. Soil types.
4. General properties of type of subsoil.
5. In situ strength and any other tests feasible on the soil exposed.
6. Samples of soil and ground water for laboratory tests.
7. Water table level.
8. Quantity of water entering the hole and description.
9. Depth of any filling on made-up ground.

From this it can be seen that shallow trial holes are of little use generally. It is therefore always advisable to take some borings down to a depth equal to the width of the building or a maximum of 30 m. This is not difficult with modern site-boring equipment.

Under an isolated pad foundation, 1.5 times the breadth is sufficient depth, for at this depth there is a safe reduced load of 20% of the foundation load (see Fig. 12a). For strip foundations it is advisable to go down to a depth three times the breadth (Fig. 12b).

Where foundations will be close together, they may act as one. They should therefore be treated as a raft, and the boring taken down to 1.5 times the width of the raft. Groups of piles are similarly dealt with, since they act or fail together (Fig. 13). Investigation of soil should be made down to a depth of 1.5 times the breadth of the group, but measured vertically down from a point two-thirds down the length of the piles (Fig. 24).

Samples of soil for testing

Samples should be carefully sealed off with paraffin wax and placed in airtight containers for transport to a testing laboratory. Additionally, porous soils should have waxed paper over the ends before sealing with wax. Each sample should be labelled with a numbered standard Sample Record Label, of which a duplicate copy is kept, recording:

1. Location.
2. Date.
3. Boring No.
4. Ordnance Survey level of ground surface at boring.
5. The position the sample was taken from, and the top and bottom levels of sample when in situ, related to Ordnance Survey level.
6. Container No. in which sample is placed.
7. Type of sample (disturbed or undisturbed).

8. Any other remarks of value for assessment, such as the presence or otherwise of ground water in the hole.
9. Signature of person taking the sample.

With samples obtained in tube samplers, it is usual to drive in the tube 150 mm, and then register the number of blows of standard force (63.5 kg dropped 760 mm) necessary to drive it in a further 300 mm. This serves as an additional guide to the relative density of the soil, and hence its strength. This is known as the Standard Penetration Test (Test 18, B.S. 1377), and the size, type of sampler and force need to be standardized by a regular experienced operator if it is to be useful (Fig. 14).

For sands, penetration tests are often made with a 90° angle cone dropped a fixed height down the borehole (see Fig. 16) or forced a fixed distance down into the soil at the bottom of the hole (Dutch cone penetrometer). The results can be indirectly correlated with shearing resistance.

Interpretation of standard penetration test

Classification of sands:

No. of blows	Relative density	Notes
0- 4	very loose	Sensitive to shock – can behave like a liquid in some conditions
4-10	loose	
10-30	medium	
30-50	dense	
Over 50	very dense	Withstands heavy load and shock

This table is not reliable in saturated fine or silty sand of medium to high density.

Consistency of clays and unconfined compressive strength

No. of blows	Consistency	Compressive strength in kN/m²
0- 2	very soft	0- 24
2- 4	soft	24- 49
4- 8	medium	49- 98
8-15	stiff	98-196
15-30	very stiff	196-392
over 30	hard	over 392

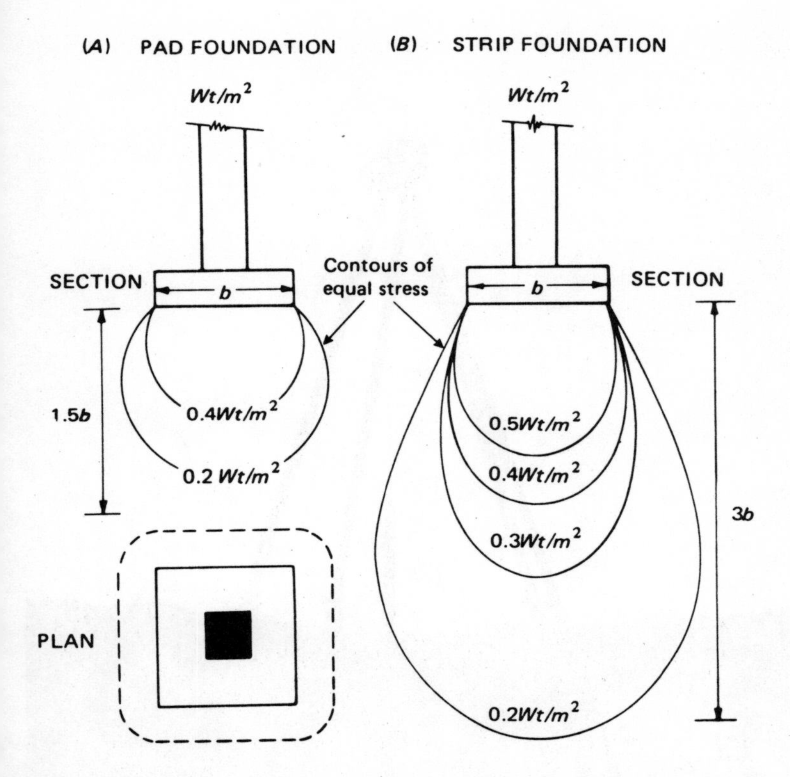

Fig. 12. Bulbs of equal pressure beneath strip and pad foundations.

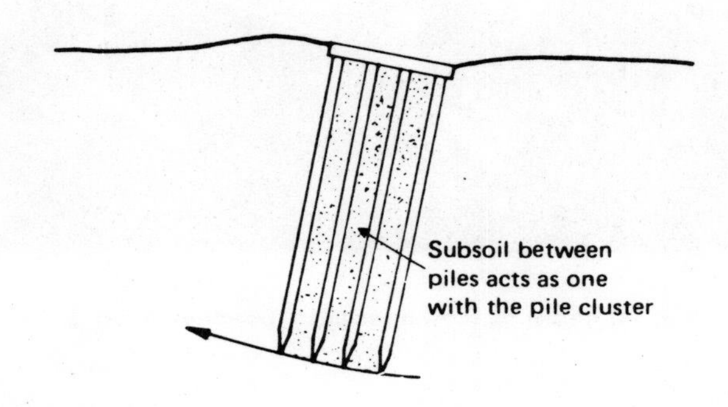

Fig. 13. Failure of pile cluster as a solid block in cohesive soil. (Rotational failure.)

Fig. 14. Boring rig showing standard penetration test in use.

Fig. 15. Examining soil sample in split-tube sampler.

Vane test for cohesive soils (Fig. 19)

This test measures the shear strength of soft cohesive soils in situ (or in laboratory samples) by forcing into the soil a rod with four equal-sized small vanes at the end and measuring the torque necessary to cause rotation. Instructions on its use are provided by the manufacturer.

Unconfined compression test for cohesive soils (Fig. 20)

This is a portable field test (Test 19, B.S. 1377) using B.R.S. developed apparatus. 38 mm dia. x 75 mm-long cylindrical samples from a trial pit or borehole can be tested to failure, which occurs by lateral bulging or diagonal shearing. The shear strength is measured without taking account of lateral pressure of soil. This is unrealistic, but the necessary allowances can be made.

Tests on non-cohesive soils

Undisturbed samples of sands and gravels are not easy to get, and where possible most tests on these soils are made in situ. Where it is essential to have an undisturbed sample of fine moist sand, this can be taken with the aid of special samplers which when pushed into the soil at the bottom of a borehole, close or make use of a piston to produce a reduced air pressure above and retain the sample. Some disturbance of density is usual, however.

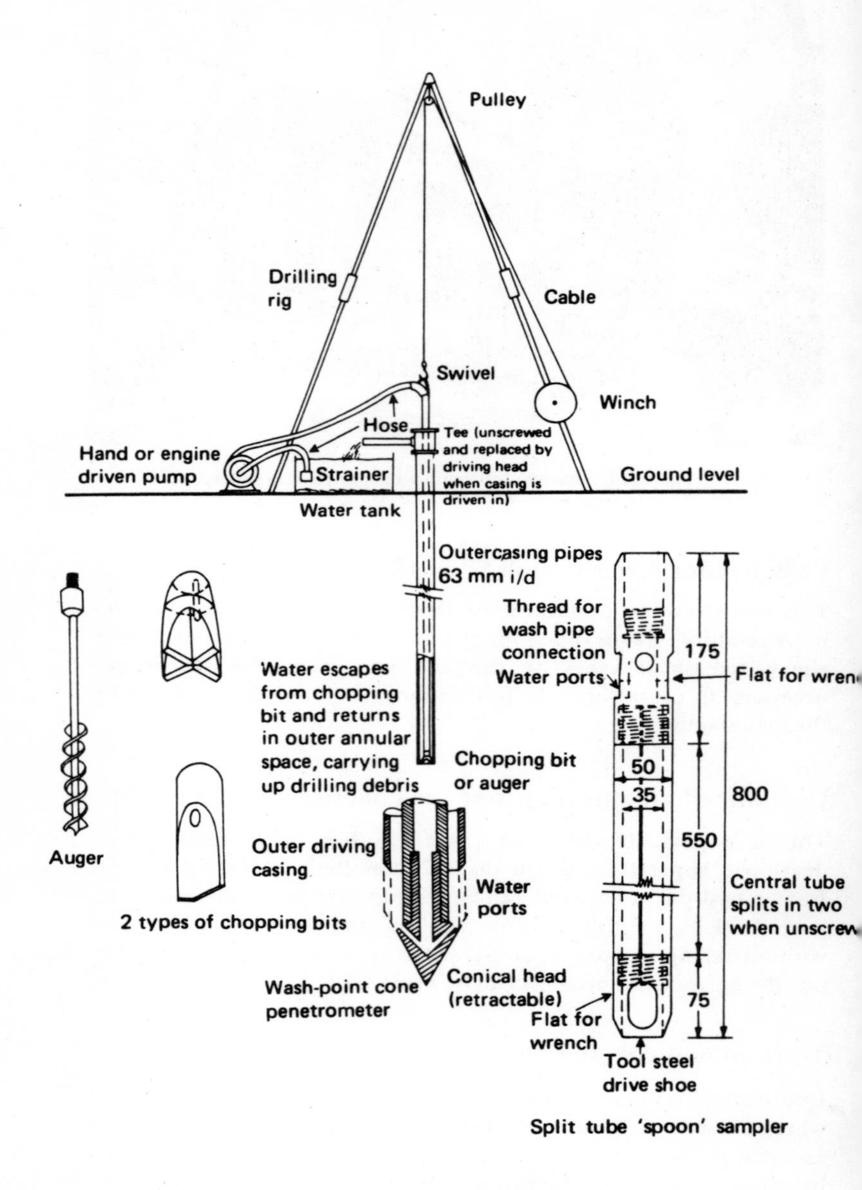

Fig. 16. Boring rig, with typical bits, subsoil sampler and cone penetrometer.

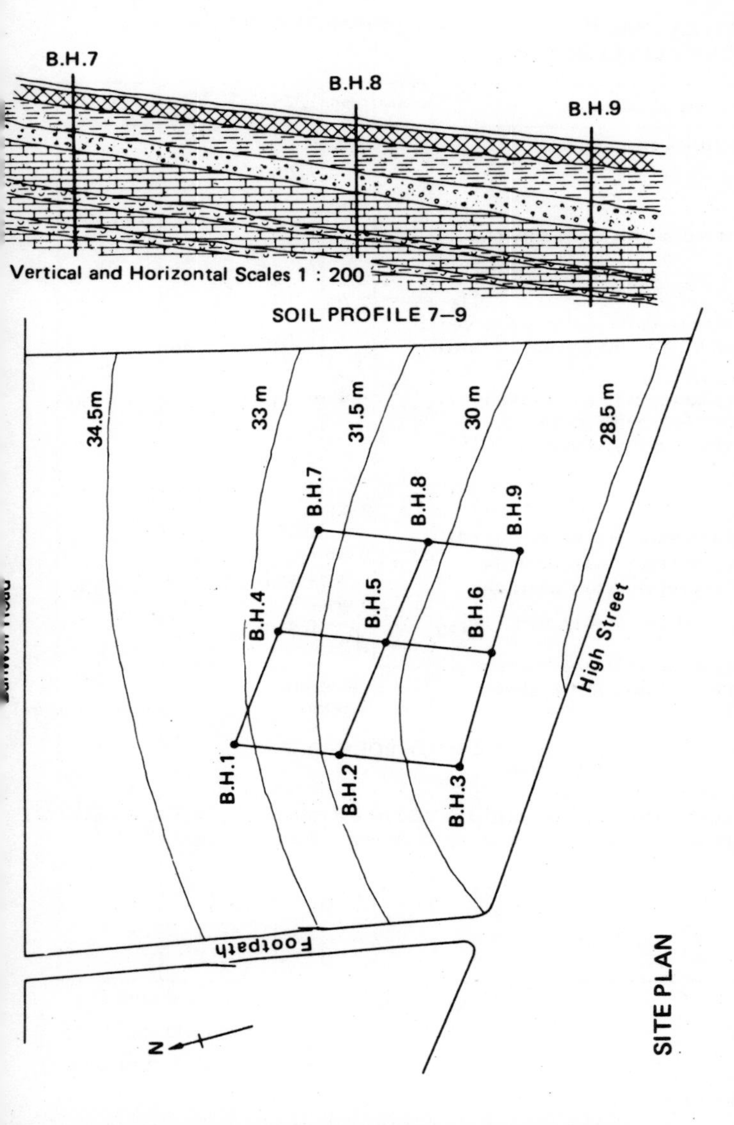

Fig. 17. Site plan, with contours, boreholes and soil profile.

BOREHOLE RECORD

BOREHOLE No. 8
GROUND LEVEL 30.36 m

DIAMETER: 150 mm
DATE:

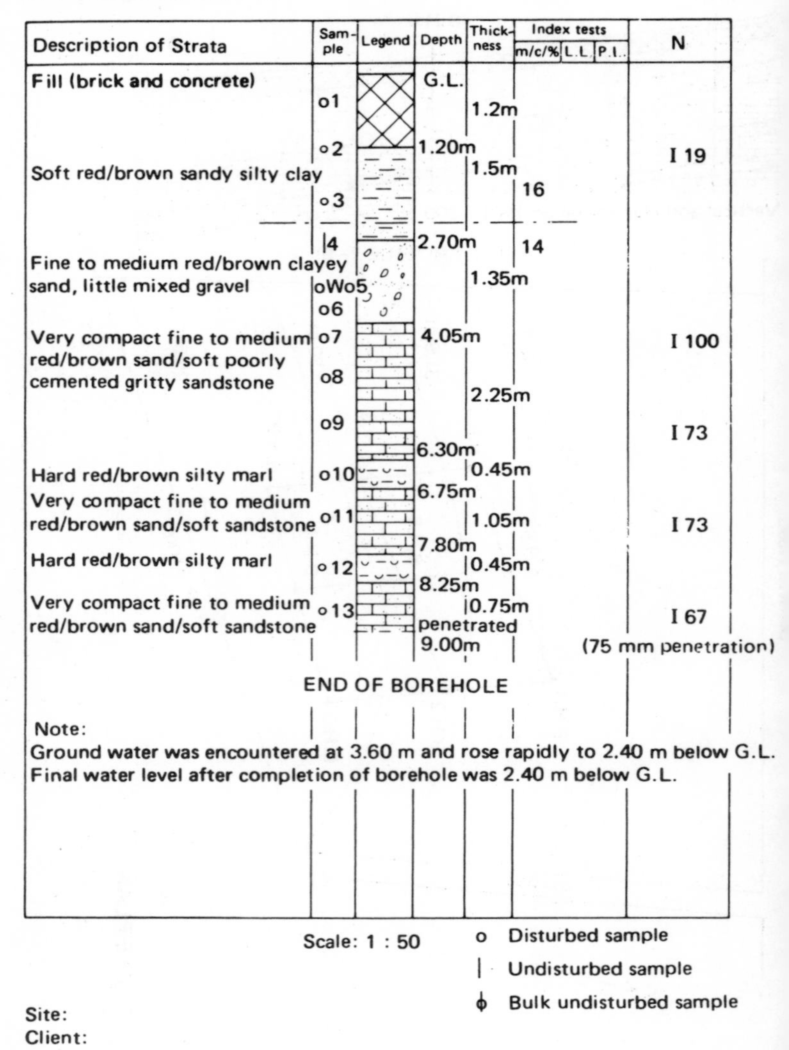

Description of Strata	Sample	Legend	Depth	Thickness	Index tests m/c/%	L.L.	P.I.	N
Fill (brick and concrete)	o1, o2		G.L.	1.2m				
Soft red/brown sandy silty clay	o3		1.20m	1.5m	16			I 19
Fine to medium red/brown clayey sand, little mixed gravel	|4, oWo5, o6		2.70m	1.35m	14			
Very compact fine to medium red/brown sand/soft poorly cemented gritty sandstone	o7, o8, o9		4.05m	2.25m				I 100, I 73
Hard red/brown silty marl	o10		6.30m	0.45m				
Very compact fine to medium red/brown sand/soft sandstone	o11		6.75m	1.05m				I 73
Hard red/brown silty marl	o12		7.80m	0.45m				
Very compact fine to medium red/brown sand/soft sandstone	o13		8.25m, penetrated 9.00m	0.75m				I 67 (75 mm penetration)

END OF BOREHOLE

Note:
Ground water was encountered at 3.60 m and rose rapidly to 2.40 m below G.L.
Final water level after completion of borehole was 2.40 m below G.L.

Scale: 1 : 50

o Disturbed sample
| Undisturbed sample
ϕ Bulk undisturbed sample

Site:
Client:

Fig. 18. Borehole record form.

Fig. 19. "Pilcon" hand vane test for shear. To carry out a test, the vane is pushed into the ground or sample until it is completely covered. The torsion head is then rotated at a rate of one revolution per minute. When the clay shears, the load on the torsion spring is released, and the pointer registers maximum deflection of the spring. The shear strength of the soil is then read off the calibration curve of the chart provided.

Fig. 20. Unconfined compression test apparatus, for use on site.

On arrival at the laboratory, split sample tubes can be unscrewed and opened for easy removal of the sample (Figs. 15 and 16) but other types of tube samplers must have the core extruded by a special tool, which once again may affect the density.

It is not always appreciated what size of samples and number are required, so the following may act as a useful guide:

For identification, moisture and chemical tests	0.5 kg of cohesive soils 4.0 kg gravels
For compaction tests	10.0 kg of cohesive soils and sands 20.0 kg gravels

For comprehensive tests on soils	20.0-40.0 kg cohesive soils and sands
	40.0-80.0 kg gravels
*For ground water samples**	1 litre bottles (corked or screwtop)

Disturbed samples of sands and silts are often kept in labelled 500 g or 1 kg standard Kilner jars (sealed with rubber rings to retain the moisture content, or in 500 g lever lid tins.

Gravels will require larger "biscuit tins" with tight lids or similar means of sealing.

Plate loading test

When a trial hole is dug to the level of the required foundation base, it is possible to carry out a *plate loading test* in which a proportionally smaller area than the real foundation is loaded by a proportionally smaller load, in order to check settlement and the safe bearing pressure on the soil. This test can be carried out in two ways. The first method is actually to stack heavy weights on a platform of timber or steel joists balanced over the test plate or a pile of known area. This is a somewhat difficult and costly project, because of the number of heavy concrete blocks, metal billets or sandbags required (Fig. 62). The second method consists of placing a hydraulic jack between the test plate and a large, heavy and immovable object (sometimes a heavy lorry or a tank of water) and pumping up the pressure registered on a gauge to the required test pressure. This latter method is generally to be preferred. Independent supports outside the "sphere of influence" of the jack hold micrometer gauges which register the settlement of the plate under load (Figs 21 and 22). Test plates are usually made of steel or cast iron, and vary in size from 300 x 300 mm on sand to 900 x 900 on clay, 600 x 600 mm being a commonly used size. Although results are more accurate with the larger sizes, the amount of kentledge (load) required is very high and the test becomes very expensive.

As an example, if an actual foundation is to measure 1200 x 1200 mm and support 32 tonnes, for a 600 x 600 mm test plate 24 tonnes would have to be applied to produce a stress of three times the required safe bearing stress of 8 tonnes force (the usual safety margin required in such tests).

As the "bulb" of equal pressure beneath the test load will be considerably smaller than the bulb beneath the actual foundation

* Bottles should be rinsed out three times with the water before final filling.

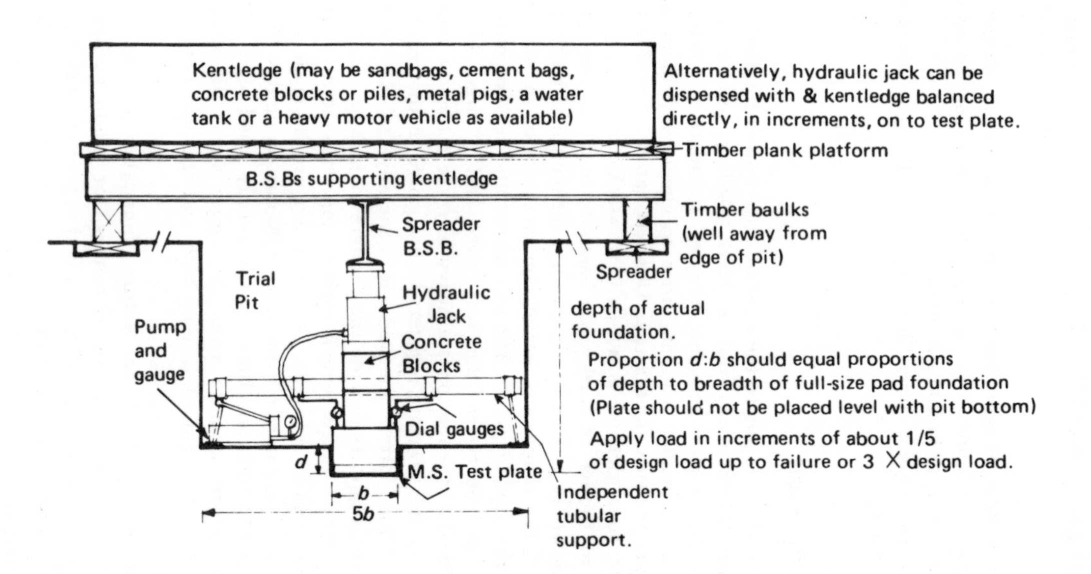
Kentledge (may be sandbags, cement bags, concrete blocks or piles, metal pigs, a water tank or a heavy motor vehicle as available)
Alternatively, hydraulic jack can be dispensed with & kentledge balanced directly, in increments, on to test plate.
Timber plank platform
B.S.Bs supporting kentledge
Timber baulks (well away from edge of pit)
Spreader
B.S.B.
Spreader
Trial Pit
Hydraulic Jack
Concrete Blocks
Pump and gauge
depth of actual foundation.
Proportion d:b should equal proportions of depth to breadth of full-size pad foundation (Plate should not be placed level with pit bottom)
Apply load in increments of about 1/5 of design load up to failure or 3 X design load.
Dial gauges
d
M.S. Test plate
b
5b
Independent tubular support.

Fig. 21. Plate bearing test.

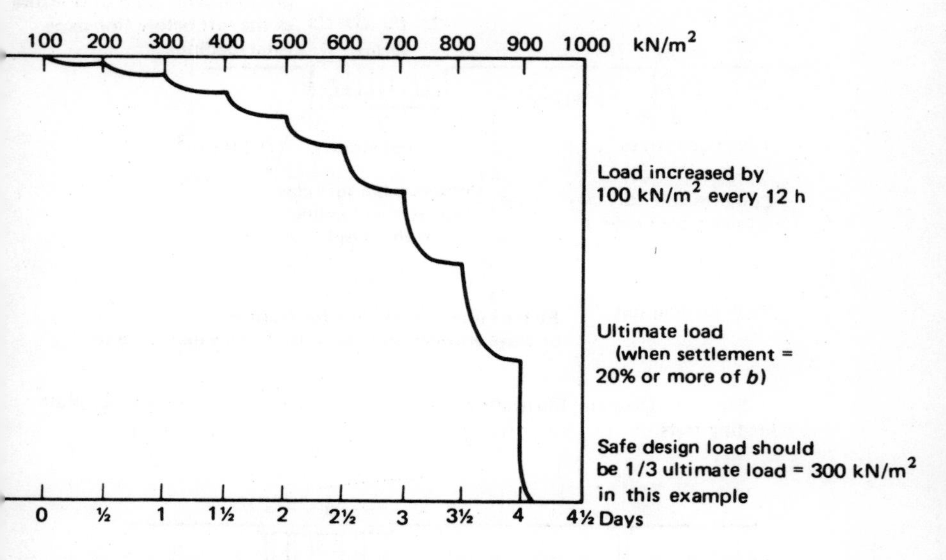

Fig. 22. Typical load-settlement graph from plate bearing test over 4½ days.

(Figs 23 and 24) there could be a failure due to underlying weaker strata which would not be revealed by this test.

Some important points to remember with the plate loading test are:

1. The test plate should be slightly sunk below the level of the bottom of the trial hole,

$$\frac{\text{depth of depression}}{\text{width of test plate}} = \frac{D}{B} \text{ of actual foundation.}$$

2. Loads are applied in increments of about one-fifth the anticipated ultimate load every 24 h, to allow time for settlement. Consequently, the test can take a long time. Settlement readings are taken every hour for the first 6 h after each weight increment is applied and then every 12 h. Loads are taken up to three times the design load if failure does not occur earlier.
3. Failure is assumed if the settlement reaches a depth equal to 20% of the diameter of the loaded area. Settlement can usually be predicted to within ± 20% of actual settlement.
4. If there is a high water table, soil must be first de-watered down to the level of the bottom of projected foundations. This is especially important with sand, or the upward water pressure may cause quicksand effect producing false readings of low strength (see later for de-watering techniques).

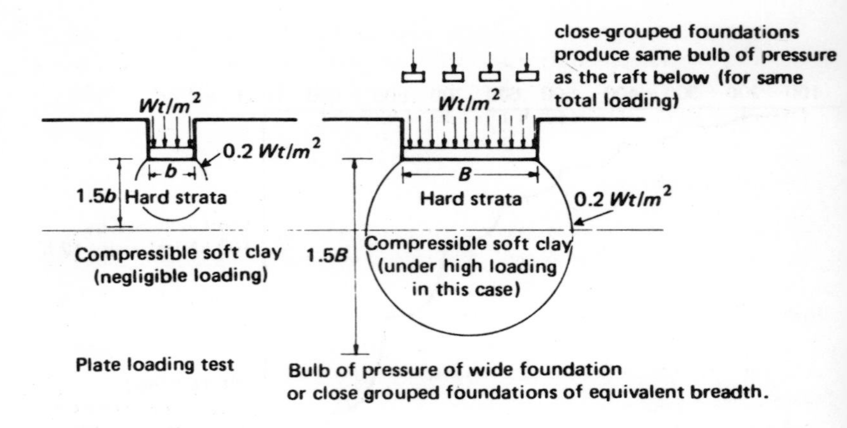

Fig. 23. Diagram illustrating the danger of relying too much on plate loading tests.

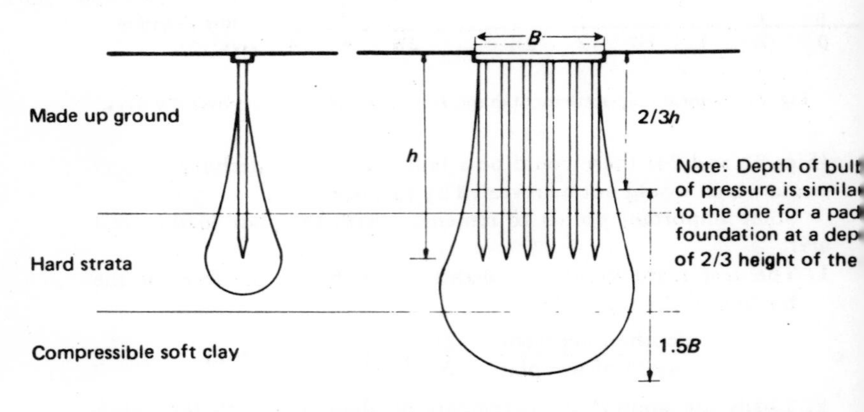

Fig. 24. Diagram comparing bulb of equal pressure resulting from loading test on a single pile with that of a group of piles.

Bearing capacity of soil

Ultimate bearing capacity (q_f) is the load at which the soil fails, or can be considered to have failed by great settlement.
Safe bearing capacity (q_s) is

$$* \frac{\text{ultimate load}}{\text{load factor}} + \text{weight etc.}$$

* Load factor is usually 3.

Allowable bearing capacity (q_a) may be *less* than the safe bearing capacity to keep settlement within certain desirable limits e.g. to avoid distortion of a building structural frame (see later), or to provide a level bed for an industrial process.

Geophysical tests

These constitute a developing field in soil mechanics, although except for special problems, their uses at the moment are fairly limited. Such tests consist of measurements of electrical resistivity, vibration recording, echo-sounding, seismic refraction and magnetic variation measurements. These and similar techniques establish changes in strata faults, the existence of cavities, the presence of metal or water, and the electrical conductivity of a soil (the last of which can be important so that steps can be taken to prevent the corrosion of steel buried in the ground).

A useful application of geophysical testing is in the realm of rock foundations which are expensive to bore through, but which can now be tested for cavities, fissures and faults with modern electronic equipment at low cost.

Chapter 4—Laboratory tests on soils

A great number of tests have been devised for the scientific analysis of soils for:

(a) *Identification and classification.*
(b) *Engineering properties.*
(c) *Chemical reactions with materials used for foundations.*

Similar tests to those in (c) are also needed when analysing ground water encountered on sites. Most of the tests involved are fully described in B.S. 1377 *Methods of Testing Soils for Civil Engineering Purposes*. Some explanatory notes on the tests, and their importance, are given below.

Identification and classification

1. Moisture content As the presence of water can affect the strength of sands and cohesive soils, it is important to know their moisture content. Test 1A is the most usual. It is interesting to compare the average m/c (moisture content) of London clay at 26% to clay under Mexico City with 242% m/c. This latter figure results in high compressibility and settlement of buildings as water is squeezed out and extracted for the local water supply. In many cases settlement has been measured at 300 mm annually and 5-6 m in 50 years.

2. Dilatency test for cohesive soils A sample of soil is shaken in the hand to produce moisture on the surface and then squeezed. If it is silt, it will dry off and powder in the hand; if clay it will not powder. (This test does not appear in B.S. 1377.)

3. Liquid limit In clays, it is often important to know whether a particular sample should be classified as solid, plastic or liquid; and at what moisture content it will change from one state to another. Tests 2, 3 and 4 of B.S. 1377 explain the standard method of determination of *Liquid Limit* (LL) using a special tilting cup apparatus (Fig. 25); of *Plastic Limit* (PL) by rolling out a pencil-like thread of plastic clay until it breaks, and then measuring the m/c; and of the *Plasticity Index* (PI), which equals LL–PL. When PL is equal to, or greater than LL, the clay is not plastic. Some silts

Fig. 25. British Standard liquid and plastic limit test apparatus.

change very rapidly from solid to liquid state e.g. PL 20%, LL 25%. For clays, the range may be 10 times this difference.

4. *Determination of linear shrinkage* (Test 5) is for cohesive soils.

5. *Specific gravity of soil particles* (Test 6).

   Part A applies to fine-grained soils

   Part B applies to medium and coarse-grained soils.

6. *Particle size distribution* (Test 7) is of particular importance for non-cohesive soils, especially if they are likely to be used for aggregates in concrete when excavated. It is also of use when there are problems of erosion and drift, and when soil stabilization is necessary. Part A of the test consists of sieving the dried sample of non-cohesive soil through B.S. 410 sieves. Cohesive soils contain microscopic particles which, in clay, are like flat plates stuck together. For these soils, wet analysis is used, employing an hydrometer or a pipette in a sedimentation technique (Parts B and C of Test 7). Comparison of the range of particle sizes is shown in Fig. 2.

Engineering properties

7. *Dry density/moisture content relationship* (Tests 11, 12, 13). These tests reveal the state of compaction of the soil, and therefore something of its load-bearing capacity. Test 14 for non-cohesive soils is better carried out on site.

8. Permeability tests are available for

(a) *Permeable soils* such as chalks, sands, gravels and silts. Sandstones should also be tested for permeability. The usual method is to measure the rate of flow of water through the soil under a hydrostatic head.

(b) *Low permeability clay*—This test uses an oedometer or consolidation press (Fig. 26) which squeezes out water from the soil under load (Test 16). This also reveals the degree of settlement probable. Squeezing out of moisture from under the middle of a large foundation raft to the perimeter has been known to cause failure on cohesive soils. The range of permeability of soils is shown in Fig. 5.

Permeability of the soil is also important for land drainage purposes, sewage disposal systems, soakaways, de-watering and soil stabilization processes.

9. Shear strength can be tested by several methods other than the ones mentioned under site testing e.g., the unconfined compression test and vane test, both of which can also be used in the laboratory.

The most usual laboratory test for shear (Test 20) is the *Triaxial test* (Figs 27, 28 and 29) which requires a detailed explanation. To simulate in situ conditions, a cylindrical undisturbed sample of cohesive soil is subjected to lateral hydraulic pressure all round and in addition a vertical load is applied on the top. Both lateral and

Fig. 26. Consolidation test equipment (oedometers).

Fig. 27. Triaxial test apparatus.

vertical pressures can be increased as required, and the bottom plate may be changed from porous to non-porous materials for different tests.

Undrained triaxial compression tests are made to give quick results using non-porous end plates. Tests are carried out at three different lateral pressures and axial stress is increased slowly until failure occurs with each sample. The results are plotted in the form of Mohr's circles. The linear envelope to the three circles gives the relationship between the principal stresses at failure. The cohesion (c) of the soil can be read off the graph as shown in Fig. 29. For cohesive soils, cohesion equals shear strength. The angle of the linear envelope to the horizontal gives the internal angle of friction of the soil. Undrained tests can also be used for wet sands.

Consolidated undrained triaxial tests are carried out after the specimen has been allowed to consolidate fully under hydrostatic pressure. It is then tested to failure under conditions of no moisture-content change.

Drained tests are those in which the axial stress is increased so slowly that no appreciable pore pressures are developed within the specimen as it is allowed to drain. The quantity of water drained off can be measured. This test is mostly used for sands but sometimes also for clay soils which have to support heavy loads.

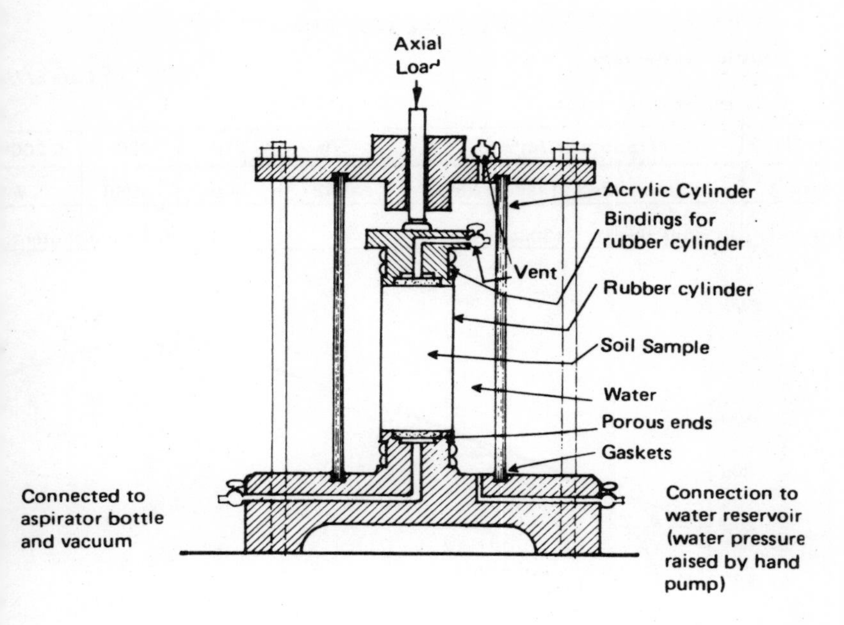

Undrained triaxial compression test results								
Borehole no.	Sample no.	Depth below G.L.	Bulk density kg/m	Moisture content %	Cell pressure m-bar	Max diff in P stress m bar	Apparent cohesion kN/m²	Ø
1	6	4.11/ 4.27	2082	12	1,000	5,040	96.51	29°
					2,000	6,900		
					4,000	11,380		
	4	2.59/ 3.05	1970	18	1,000	2,690	120.66	9°
					2,000	3,520		
					4,000	4,450		

Fig. 28. Section of triaxial test apparatus with examples of results of tests.

Sands are often tested in a shear box apparatus (Figs 30 and 31) in which the sample (a rectangular prism of sand) is subjected to an axial load normal to the direction of shear force being applied until it fails. For drained tests, water escapes through porous stones and perforated grilles. For undrained tests, solid metal grilles are used but these tests are usually only used for soils of low permeability. An example of the results of a shear box test and its interpretation is given in Fig. 31.

TRIAXIAL TEST RESULTS

Client Mardale and Wantage

Date 5/7/

Site Backwell Re-Development

Borehole No. 4	Cell Pressure kN/m^2	70	210	420	C 200
Sample No. 1	Compr. stress at failure kN/m^2	648	800	1046	Ø
Depth in m 1	Bulk density 2130kg/m^3		Moisture content		

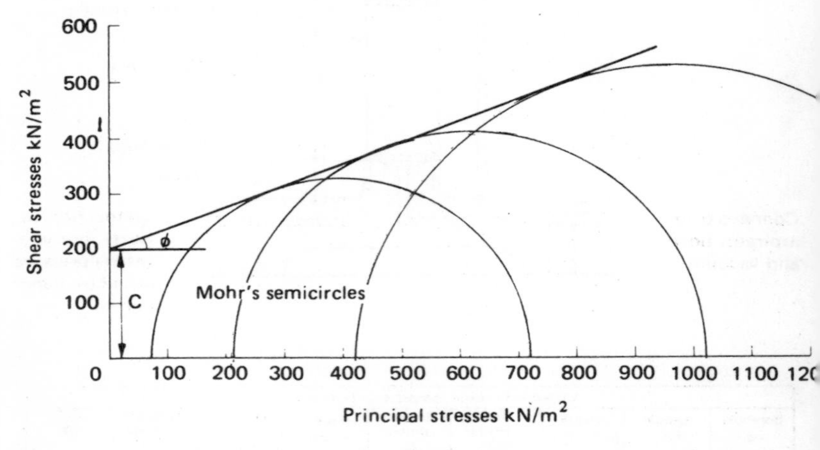

Example of triaxial test report on a stiff clay

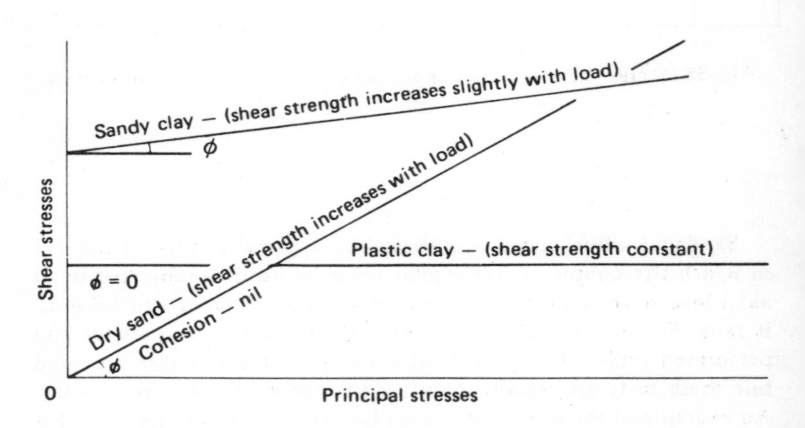

Fig. 29. Examples of results of triaxial tests with different types of soils.

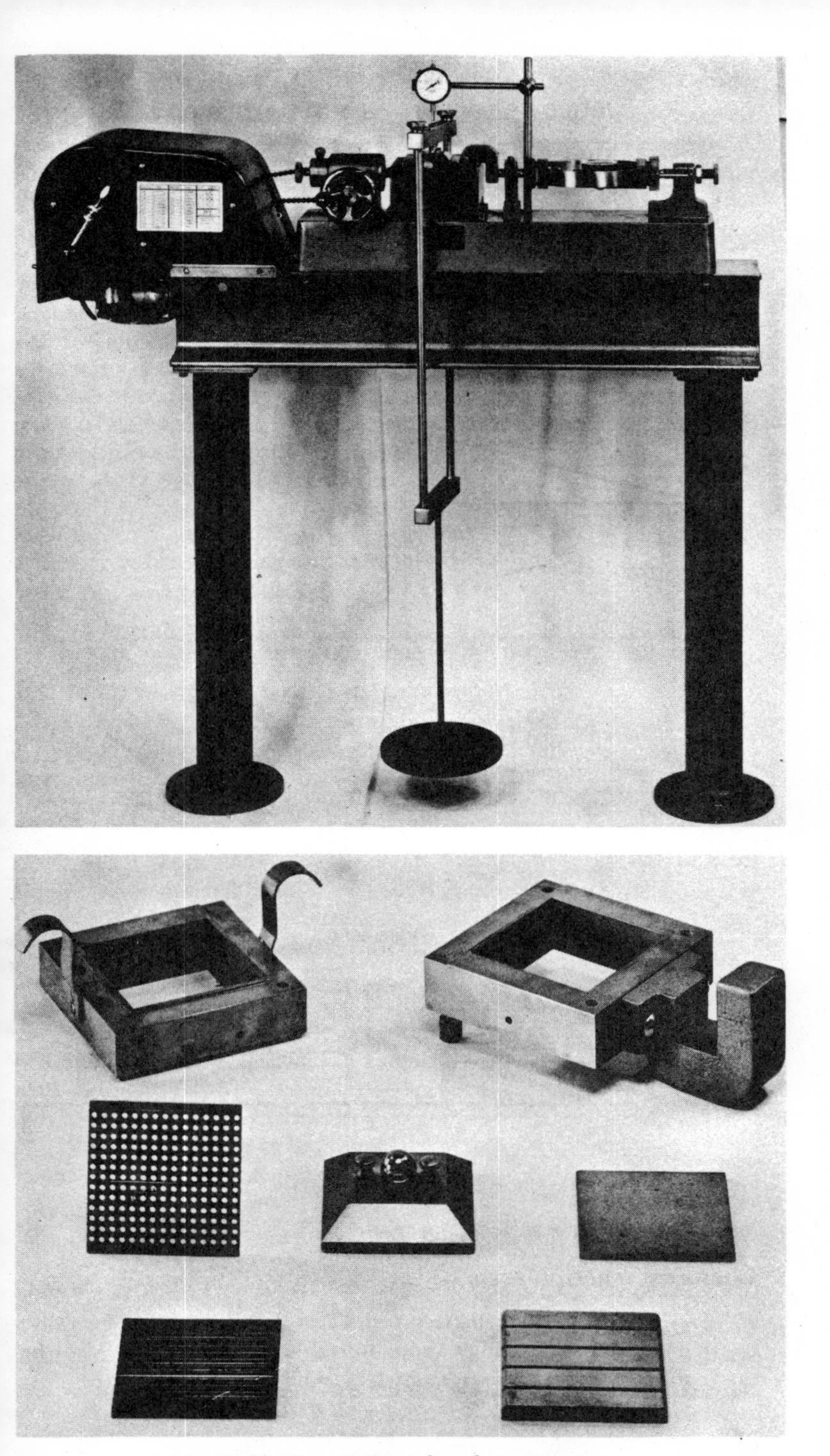

Fig. 30. Multispeed direct-shear box apparatus.

RECORD OF TEST WITH 60 x 60 mm TEST BOX

Normal Load (newtons)	248	494	745
Shear Load (newtons)	298	359	421

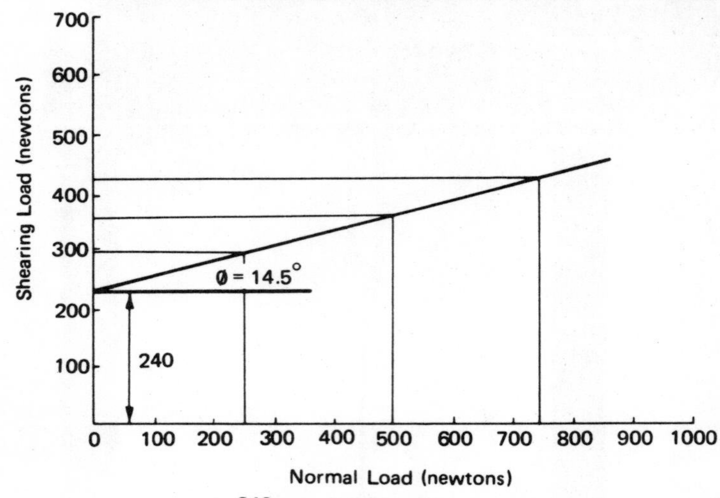

$$S = \frac{240}{360} = 0.667\ \text{N/mm}^2 \text{ at zero normal stress}$$

SHEAR BOX TEST RESULTS GIVING
ANGLE OF INTERNAL FRICTION & SHEAR STRESS

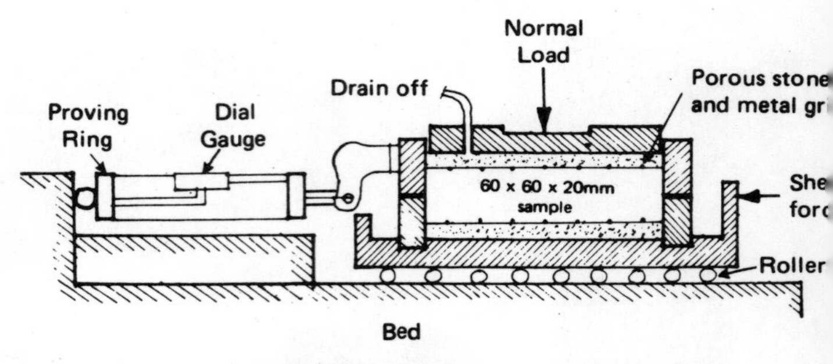

Fig. 31. Shear box test.

Chemical reaction tests

10. Presence of organic matter test (Test 8 B.S. 1377). Generally, organic matter is spongy and unstable, and it may also be unsatisfactory in an aggregate which is to be used for concrete.

11. Presence of sulphates test (Test 9) can be used for both soil or ground water. Sulphates are usually calcium, sodium or magnesium. There are tests also for the presence of sulphate-reducing bacteria. Sulphates cause deterioration of concrete and unless it is protected, corrosion of steel. Sulphate-resistant cement can be used if the danger exists. Table 4 explains when this precaution is necessary.
12. Acidity test (Test 10) for both soil and ground water is advisable as corrosion can occur in steel if the pH value is low (as in peat). Rainwater is usually acid, especially in cities, down to pH 3.9.*

A site investigation report will contain the results of many such tests. It will also give recommendations on suitable bearing pressures on the soil and the depth for foundations in the soil below G.L. at points where boreholes were driven.

Further references on soil investigation, soils and soil mechanics

B.S. 1377 Methods of testing soils for civil engineering purposes — B.S.I.
B.S. 722 Borehole and well pump tests — B.S.I.
B.S. Code of Practice 2001 – Site investigation — B.S.I.
B.S. Code of Practice 2003 – Earthworks — B.S.I.
The mechanics of engineering soils – P. Capper and W. Fisher Cassie — Spon
Soil Mechanics in engineering practice – K. Terzaghi and R. Peck — Wiley, N.Y.
Foundations and soil properties – R. Hammond — Macdonald
Geology for Engineers – F. Blyth — Arnold
The Building Regulations 1972 — H.M.S.O.
B.R.S. Digests (1st series) — H.M.S.O.
No. 3 House foundations on shrinkable clay
No. 9 Building on made-up ground or filling
B.R.S. Digests (2nd series) — H.M.S.O.
Nos. 63, 64, 67 Soils and Foundations Parts 1, 2, 3
No. 75 Cracking in Buildings
Influence charts for computation of stresses in elastic foundations – Newmark — University of Illinois Engineering Experimental Station
Laboratory testing in soil engineering – T. N. Ackroyd — Soil Mechanics Ltd.

*pH values are measured from 0-14; 0 is very acid, 14 very alkaline, 7 neutral.

Table 4. Classification of sulphate soil conditions affecting concrete, and recommended precautionary measures

Classification of soil conditions			Precautionary measures		
				Cast in situ concrete	
Class	Sulphur trioxide in ground water (Parts SO_3 per 100 000)	Sulphur trioxide in clay (per cent SO_3)	Pre-cast concrete products	Buried concrete surrounded by clay	Concrete exposed to one-sided water pressure, or concrete of thin section
1	Less than 30	Less than 0.2	No special measures	No special measures, except that the use of lean concretes (e.g. 1 : 7, or leaner, ballast concrete) is inadvisable if SO_3 in water exceeds about 20 parts per 100,000. Where the latter is the case, Portland cement mixes not leaner than 1 : 2 : 4, or, if special precautions are desired, pozzolanic cement or sulphate-resisting Portland cement mixes not	No special measures, except that when SO_3 in water is above 20 parts per 100,000, special care should be taken to ensure the use of high quality Portland cement concrete, if necessary 1 : 1½ : 3 mixes; alternatively, pozzolanic cements or sulphate-resisting Portland cement may be used in mixes not leaner than 1 : 2 : 4.

			are not likely to suffer seriously, except over a very long period of years. Alternatively, either pozzolanic, sulphate-resisting Portland, high-alumina or supersulphate cement should be used.	are not likely to suffer seriously over a short period of years, provided that care is taken to ensure that a very dense and homogeneous mass is obtained. For most work, and particularly if the predominant salts are magnesium or sodium sulphates, concrete made with either pozzolanic cement, sulphate-resisting Portland cement, high-alumina cement or super-sulphate cement (1 : 2 : 4) is advisable. See Note 1.	Pozzolanic cement or sulphate-resisting Portland cement or, preferably, either high-alumina cement or supersulphate cement is recommended.
3	Above 100	Above 0.5	The densest Portland cement concrete is not likely to suffer seriously over periods up to, say, 10-20 years, unless conditions are very severe. Alternatively, high-alumina or supersulphate cement concretes should be used.	The use of high-alumina or supersulphate cement concretes is recommended.	The use of high-alumina or supersulphate cement concretes is recommended.

NOTES: 1. Where 1 :2 : 4 concrete is mentioned, other mixes of equivalent weight ratio of cement to total aggregate, but with somewhat increased ratio of sand to coarse aggregates (e.g. $1 : 2\frac{1}{4} : 3\frac{3}{4}$, or even $1 : 2\frac{1}{2} : 3\frac{1}{2}$), may be used, sometimes with advantage. It may be necessary when using supersulphate cement to employ mixes somewhat richer than 1 : 2 : 4 in order to obtain adequate workability.

2. Adequate assurance should be obtained that cements claimed to be sulphate-resisting Portland cements have, in fact, a high resistance to sulphates. Where the whole of the work, under adverse conditions, cannot be done with resistant cements, protection should be given either by casing with a layer of resistant-cement concrete or by coating with bituminous materials.

(Table 4, C.P.3—Ch. IX Durability—B.S.I.)

SECTION TWO

The foundation

5. Principles of foundations
6. Selections of foundation type
7. Piles and piling
8. Retaining walls
9. Some special problems in foundation design
10. Preparing the site for foundations
11. Materials for foundations

Chapter 5—The principles of foundations

Having discovered the nature of the site, its contours and surface features; what lies under it; its rock and soil types; and finally having received the results of tests on the bearing capacity and chemical reactions of the soil, the designer is at last in a position to proceed with the design of the foundations for his building.

Decisions now have to be made, but they can only be made with confidence when the designer understands the purpose of the foundation; can estimate the loads on the building's structure, the magnitude and direction of forces applied to each foundation; and is aware of the range of alternative types of foundation available.

The purpose of foundations

Foundations must be designed:

1. To intercept loads and tension forces from the building structure, such as from walls, columns and ties, and divert and spread them over a large enough area to utilize the maximum allowable resistance of the soil. This will have been pre-determined and will allow for possible errors or weaknesses by the use of a factor of safety. It will also provide for an acceptable amount of settlement or "give" which will not induce too great a strain in the structure, or in the case of a tension structure, allow too much "sag". It is not always compressional forces which the soil and foundation may have to resist but also tensional and horizontal forces involving friction and adhesion. The ability to analyse, and visualize these forces in a state of movement, pushing and pulling at the soil and foundation, is vital to the designer. Whenever possible, it is preferable for the load to be placed concentrically on a foundation, or in a balanced way, as eccentric loading induces bending moments which subject the ground to heavier loading on one side (Figs 32c and 74).
2. To be strong enough to prevent downward vertical loads shearing through the foundation itself at the point of application (Fig. 32b shows a typical shear failure).

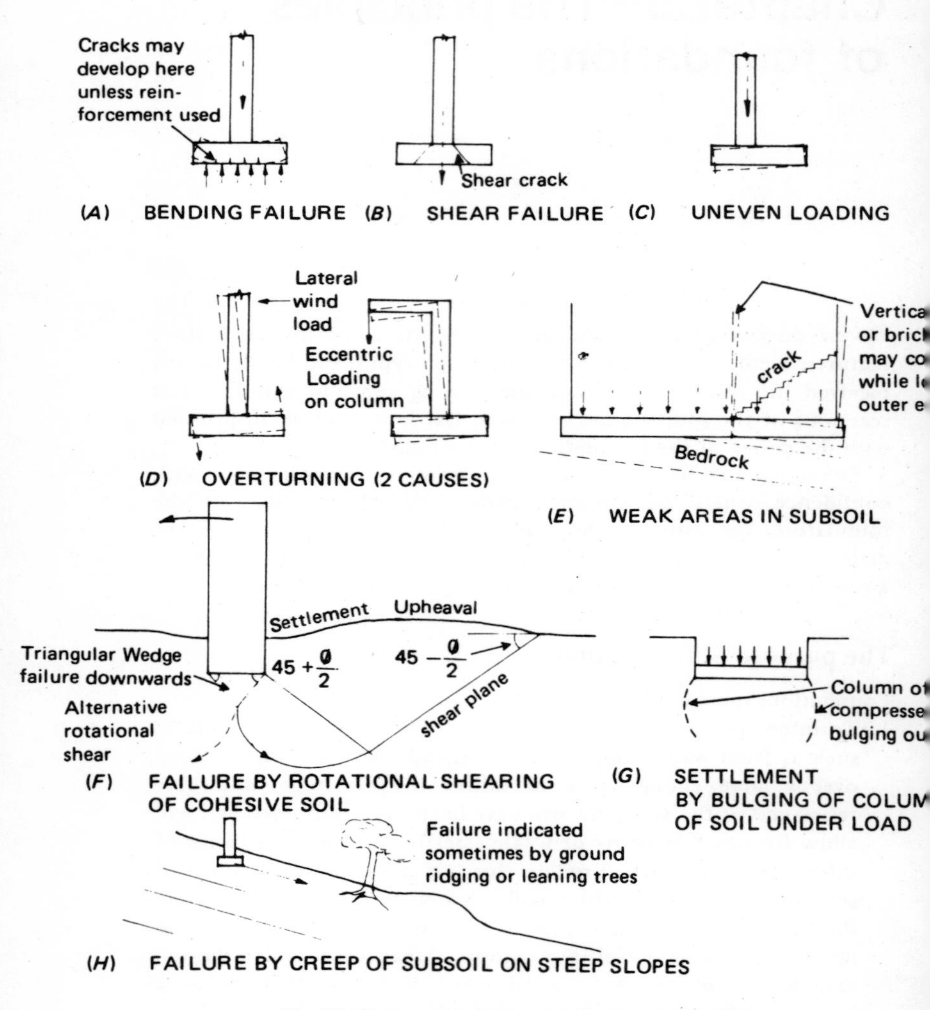

Fig. 32. Some common reasons for foundation failure.

3. To withstand tensional and shearing forces generated by the tendency of the foundation to bend under the opposing forces of a concentrated load from the structure above and the distributed resistance from the soil under (Fig. 32a). By visualizing a flexible foundation bending under these applied forces, it becomes clear where tensional reinforcement will be necessary in the concrete,

which (unless it is very thick) will not have enough tensional strength in itself. Except for lightly loaded foundations, it is usually more economical to use concrete reinforced with steel to resist tension, than to use concrete of sufficient thickness to withstand the tension and shear forces induced in it.

4. To avoid utilizing unreliable or weak ground by carrying loads down to deeper sound rock or soil in some instances (Fig. 32e).
5. To utilize buoyancy when founded on plastic or liquid soils of too great a depth to use the principle in (4) above.
6. To accommodate movements of the ground due to swelling and shrinkage, unstable ground, mining activities (Fig. 33) or the horizontal forces of earthquakes, all of which may alter the stresses within the foundation, requiring considerable forethought on the disposition of reinforcement, or on the type of foundation to be used.
7. To withstand attack from corrosive elements in the soil, or water in contact with them, and in some cases to withstand organic attack from fungi, bacteria or insects.
8. To withstand water pressure where this may occur.

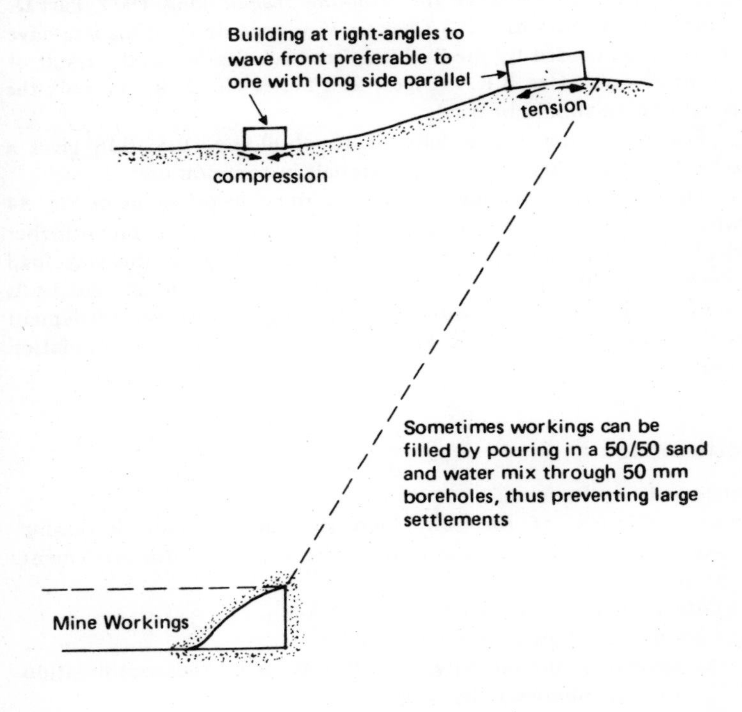

Fig. 33. Effects of mining subsidence.

In addition to these basic requirements, the foundations must also be designed in accordance with the statutory requirements of the Building Regulations and meet with the approval of the local Building Inspector, or in London, the District Surveyor.

It is salutory to bear in mind that in some South American countries, the penalty for incompetence in the design of a building, resulting in its collapse, is imprisonment for a long stretch! If the cost of a building is to be cut, one should certainly never economize on foundations.

Assessing the load on foundations

The load imparted by the building on a particular foundation under examination, must be estimated. This implies a careful analysis of dead loads and live loads including any moving loads, such as cranes or vehicles, when the building is in use. C.P.3 Chapter V of the *Code of Functional Requirements of Buildings* issued by the B.S.I. suggests minimum figures for calculating loading. Statutory minimum figures are given in the *Building Regulations* 1972 Part D. Some amendments to C.P.3 Chapter V as regards wind loading have been recommended by the Building Research Station as the result of recent research (see B.R.S. Digest 2nd Series No. 119) and only the latest edition should be used.

For estimating the weight of the building, B.S. 648 gives a schedule of weights of building materials in common use.

The analysis of loading can provide some surprises, as in Fig. 34 where the load on the central partition foundation is much higher than at first seems apparent. An understanding of the way load forces are diverted and concentrated by openings in walls (sometimes resulting in unbalanced loading, and uneven settlement with cracks forming as in Fig. 35) can lead to better foundation design.

Settlement

Settlement results from:

1. *Consolidation of the soil,* increasing density and decreasing volume. This can increase load-bearing capacity until settlement stops.
2. *Lateral bulging of cohesive soils* (Fig. 32g).
3. *Elimination of water* in the soil by:
   (a) *squeezing out* laterally, as with clay or by the consolidation of non-cohesive soils;
   (b) *drying out* sometimes a seasonal occurrence; at other times it may be caused by heating boilers or by tree roots;

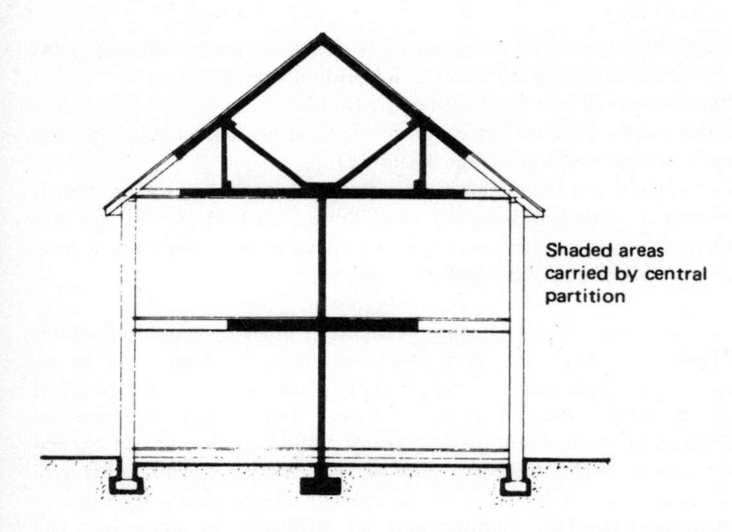

Fig. 34. Internal load-bearing partitions may carry greater loading than external walls and therefore need larger foundations.

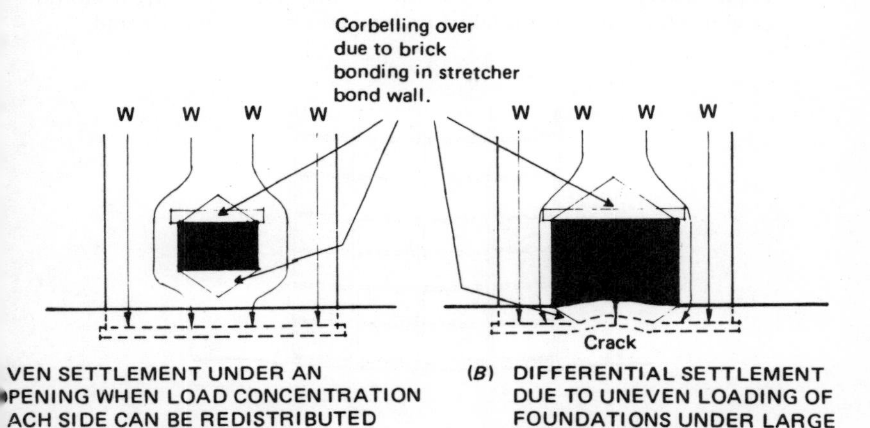

Fig. 35. Importance of even load distribution, and planned even settlement.

(c) *pumping out*, as when water is extracted for water supply (as in Mexico City) or to keep a building site dry,

4. *Plastic flow* of soil from under the building.
5. *Mining activities*, or the collapse of underground natural cavities such as swallow holes (Figs 3 and 33).
6. *Erosion of soil* from under foundations by wind or water. This is common with buildings on sand dunes, and where springs and land drainage water carry away soil in suspension over a long period, particularly behind retaining walls.

Except on solid rock, some settlement *must* be expected under the load of a building. It is only necessary to limit this to an acceptable amount, and to ensure that it will occur evenly to avoid straining or cracking a frame or wall (Fig. 32e). Differential settlement between columns in a framed building should not exceed 20 mm, and deflection/span should not exceed 1/300. This will avoid inducing too great a strain on the frame joints by angular distortion (Fig. 36). Frames can be stiffened to even out the differing stresses, but this induces high moments in the frame members.

Differential settlement of flexible foundation rafts (see later) should be limited to 1/500 of their width. This gives a reasonable factor of safety, as damage usually occurs only when deflection is greater than 1/300 of the width.

Even if a structure can accept differential settlement, it is probable that the plaster on walls will crack. Drainage falls under a building may also be reduced by settlement, and some margin should be allowed for this. Roof falls to rainwater may also be affected.

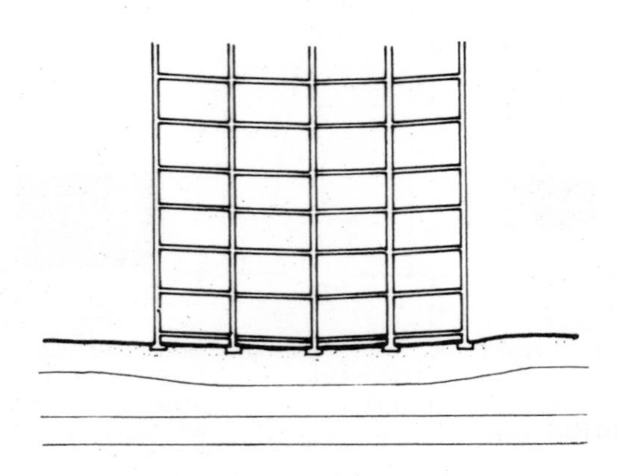

Fig. 36. Differential settlement due to compressible substratum producing strain on frame joints, out-of-level floors, distortion of infill panels.

Where movements are likely to be great, as in mining areas, it is possible to incorporate jacks into the foundations which can be used to compensate settlement at intervals of time.

If designed for, the settlement of a tall or heavy building can be quite large, and 150 mm is not unusual. To ensure even settlement, structures such as tower blocks should be symmetrical. Special care is needed when a low podium is planned around the base of a tall tower block (Figs 37 and 38).

Settlement of buildings on clay soils has already been mentioned when describing the characteristics of clay in Chapter 2.

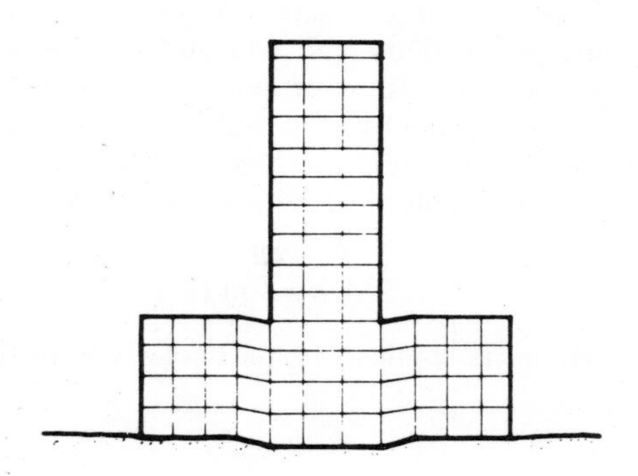

Fig. 37. Differential settlement due to uneven loading of subsoil.

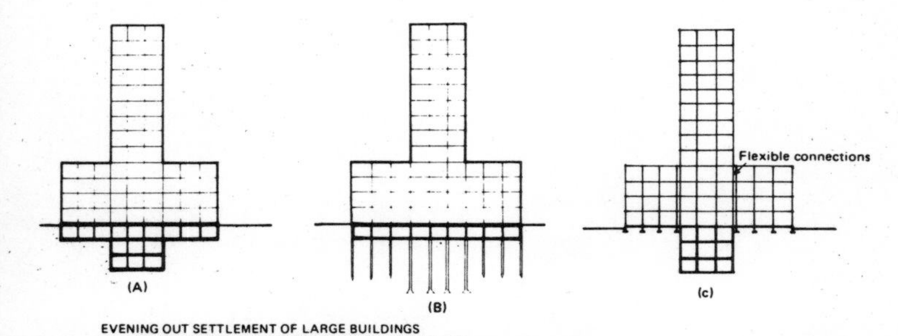

Flexible connections
(A)
(B)
(c)
EVENING OUT SETTLEMENT OF LARGE BUILDINGS
A) BY USING DEEPER BASEMENT UNDER TOWER BLOCK
B) BY USING FRICTION PILES OF GREATER DEPTH & WITH ENLARGED ENDS UNDER TOWER
NOTE: WITH ROCK AT A SUITABLE DEPTH, END-BEARING PILES COULD BE USED.
C) BY CARRYING THE TOWER DOWN ON FRAME & FOUNDATIONS DIVORCED FROM THOSE OF THE PODIUM.

Fig. 38. "Evening-out" the settlement of large buildings.

Prediction of probable settlement

There are no very accurate methods of predicting settlement. Knowledge of local geological conditions and measurements of the settlement of other buildings in the area can be a useful guide.

A plate loading test will provide measurable settlement and this information can be used to assess the probable downward movement of larger or more heavily-loaded foundations. With the same load per unit area, comparative settlement of square pad foundations is roughly proportional to the comparative breadth of each with cohesive soils. With non-cohesive soils, the settlement is usually less and not always proportional to the breadth of the foundations. It is best to use all available methods and information before hazarding an estimate. Tests of soil characteristics such as bulk density, moisture content, permeability and consolidation tests build up a soil profile. Water table level is important.

A formula developed by Terzaghi and Peck relates a 300 mm x 300 mm plate test settlement Δ_1, to the actual settlement Δ of a foundation of breadth B mm, thus:

$$\Delta = \Delta_1 \left(\frac{2B}{B + 300} \right)^2$$

Caution is needed in its application, particularly with non-cohesive soils.

Chapter 6—Selection of foundation type

Although there are quite a number of different types of foundation, most are applicable to particular situations, and there is rarely any choice once all the parameters are known. The main types of foundation are as follows:

1. Strip foundations
2. Pad foundations
3. Raft foundations
4. Buoyancy foundations
5. Piled foundations

A description of each and indication of its general application is given below.

Strip foundations

A continuous strip of concrete or reinforced concrete rests on the soil at a depth and of a width dependent on the type of soil (Figs 39, 40 and 41). Such foundations are used to support load-bearing walls in cell and cross-wall type construction on average to good bearing capacity soils. They are also used for boundary walls, masonry retaining walls, and partition walls of high self-weight in single-storey buildings. It is inadvisable to use strip foundations for the infill ground floor walls of framed buildings, as they may settle or move independently of the frame foundations. Neither are they suitable on very soft clay, silt or peat; or on badly made-up ground. Sometimes, r.c. strip foundations are used for a row of closely-spaced columns, as it is often cheaper to excavate a foundation trench mechanically than to dig a series of isolated rectangular holes.

For buildings where the total load on the foundations does not exceed 66 kN/m run, and there is no undue concentration of load, the width of strip foundation should be not less than:

1. That shown in the table to Regulation D7 of the *Building Regulations* 1972 (Table 2) or
2. the overall thickness of the base of the wall, whichever is the greater.

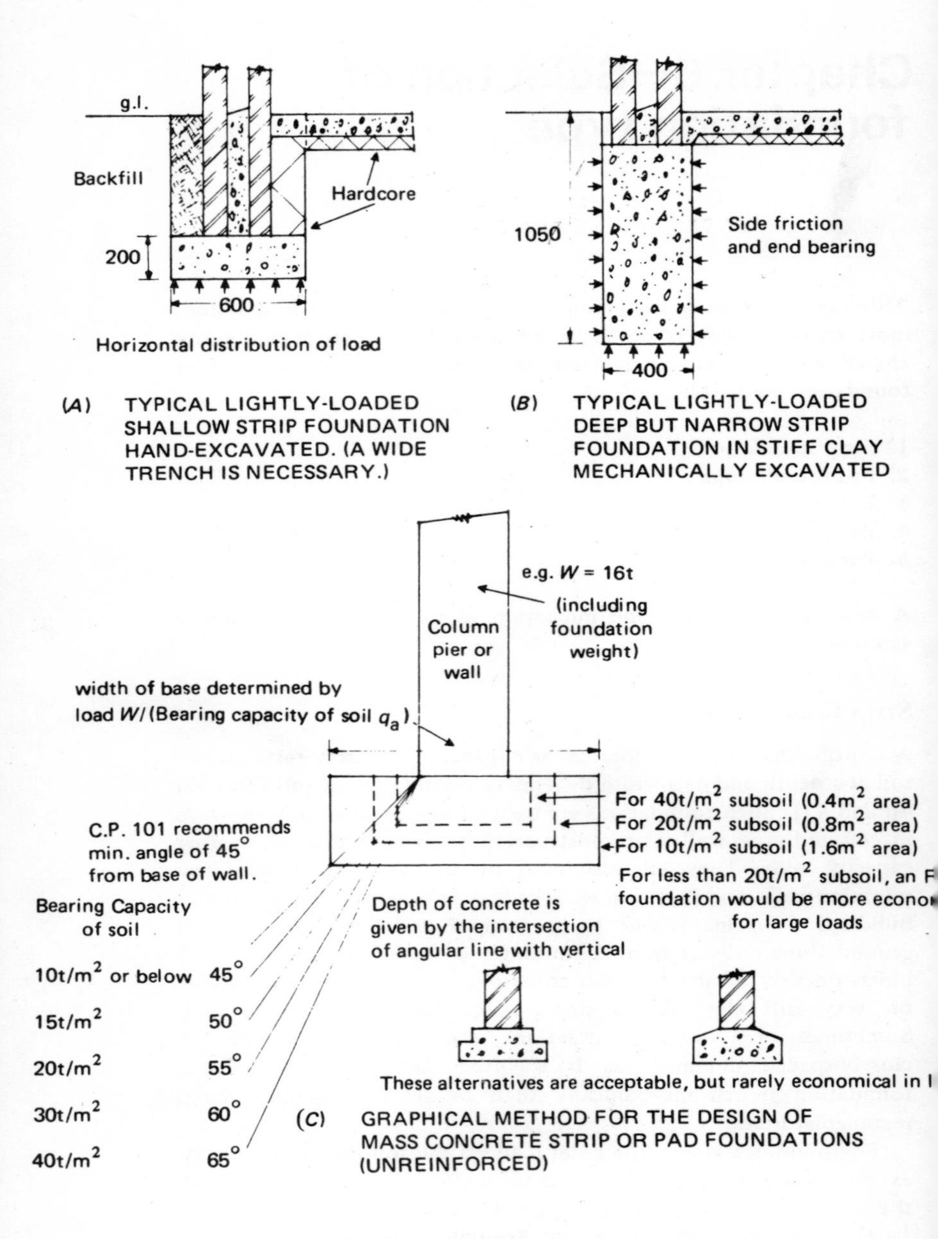
g.l.
Backfill
Hardcore
200
600
Horizontal distribution of load
(A) TYPICAL LIGHTLY-LOADED SHALLOW STRIP FOUNDATION HAND-EXCAVATED. (A WIDE TRENCH IS NECESSARY.)
1050
Side friction
and end bearing
400
(B) TYPICAL LIGHTLY-LOADED DEEP BUT NARROW STRIP FOUNDATION IN STIFF CLAY MECHANICALLY EXCAVATED
e.g. W = 16t
(including
foundation
weight)
Column
pier or
wall
width of base determined by
load W/(Bearing capacity of soil q_a)
For 40t/m² subsoil (0.4m² area)
For 20t/m² subsoil (0.8m² area)
For 10t/m² subsoil (1.6m² area)
C.P. 101 recommends
min. angle of 45°
from base of wall.
For less than 20t/m² subsoil, an F
foundation would be more econo
for large loads
Bearing Capacity
of soil
Depth of concrete is
given by the intersection
of angular line with vertical
10t/m² or below 45°
15t/m² 50°
20t/m² 55°
30t/m² 60°
40t/m² 65°
These alternatives are acceptable, but rarely economical in l
(C) GRAPHICAL METHOD FOR THE DESIGN OF MASS CONCRETE STRIP OR PAD FOUNDATIONS (UNREINFORCED)

Fig. 39. Simple foundations of unreinforced concrete.

The thickness of strip foundations in which no transverse reinforcement is provided should be not less than:

1. 150 mm, or
2. the distance the foundation projects from the base of the wall, or footing, as the case may be, whichever is the greater;
3. in the case of sleeper walls (supporting only suspended timber ground floors) 75 mm;
4. in the case of partition walls not more than one storey in height, 100 mm.

It must be emphasized that the above are *minimum* sizes and may not necessarily be sufficient for the loads they have to carry.

The footings mentioned above are rarely used nowadays, due to the better quality concrete now commonly used for foundation slabs.

It is important that foundations are centred under the wall they support to avoid the heavier loading on one edge resulting from eccentricity (Fig. 32c). Where a pier or buttress forms part of a wall, the foundation must project round on all sides at least to the same extent as it projects beyond the base of the wall.

It is recommended that longitudinal reinforcement bars should be used wherever there is an abrupt change in magnitude of load or variation in ground support, or a big change in level. It is difficult to estimate exact requirements, but two 10 or 12 mm bars at the top and two at the bottom of the foundation of walls up to 300 mm thick should be sufficient.

Properly calculated reinforced concrete strip foundations (Fig. 41) should be used where the load needs to be spread over a wide area, and with heavy loading, when the volume of mass concrete becomes excessive, and therefore uneconomic (see appendix for calculation method).

Pad foundations

These are usually square on plan (Fig. 42), but may be rectangular near boundaries of a site, or circular if the excavation for them is bored by auger. They can be formed of mass concrete, but the larger ones are usually of reinforced concrete in order to spread the heavier loads over a large area. They are generally used to support columns and piers, or heavy machinery in a factory. Pre-cast concrete columns can be wedged into holes formed in the pad (Fig. 43a), or bolted down, as is usual with a pin-jointed steel column (Fig. 43b) with bolts built into the foundation when it is cast, arranged to allow some tolerance for positioning. Pin joints for both steel and concrete columns are shown in Fig. 44 and both pin and fixed end conditions for steel columns in Fig. 45. For in situ r.c. columns,

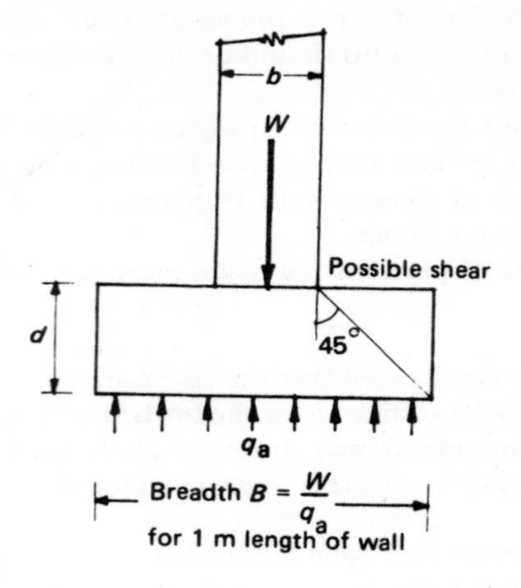

Note: A cavity wall should be drawn as a solid wall, with B = sum of thicknesses of both leaves to find required depth d

Fig. 40. Strip foundation (unreinforced concrete).

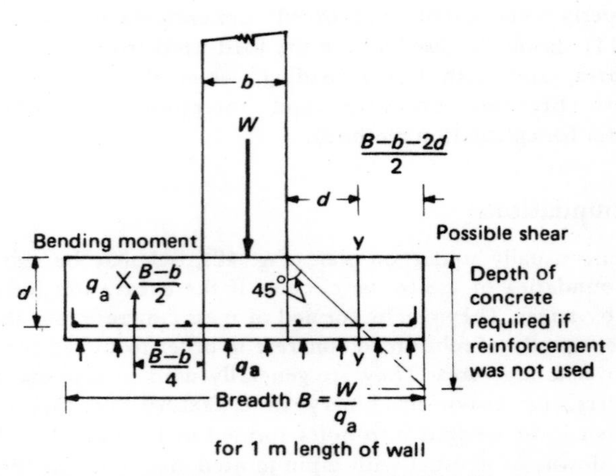

Moment taken about face of wall (C.P.114)

Fig. 41. Strip foundation (reinforced concrete).
(See Appendix, page 150.)

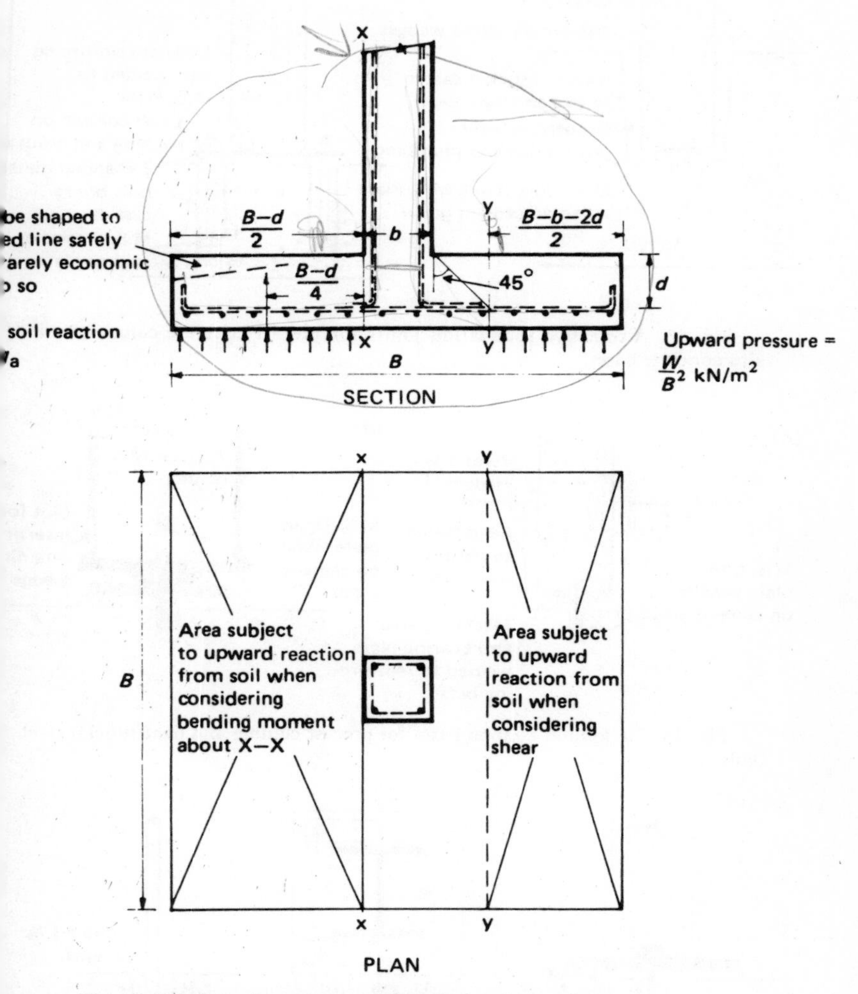

Fig. 42. Reinforced concrete pad foundation for a column.
(See Appendix, page 153.)

starter bars are tied in to the reinforcement of the base and left projecting upwards. Timber portal frames or columns are best supported by an intermediary metal shoe, dowelled into the foundation (Fig. 46).

The required area of a pad foundation is obtained by dividing the building load by the bearing capacity of the soil, allowing of course for the weight of the foundation itself. For example, a total load of 360 kN on clay of 200 kN/m^2 allowable bearing capacity, will

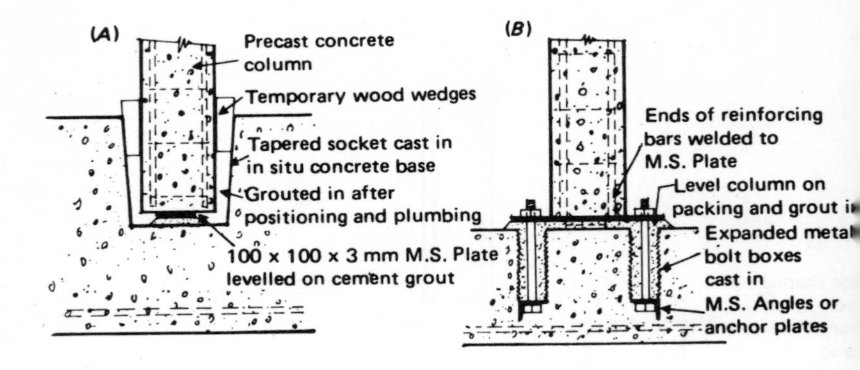

Fig. 43. Alternative foundation joints for precast concrete columns to in situ concrete bases.

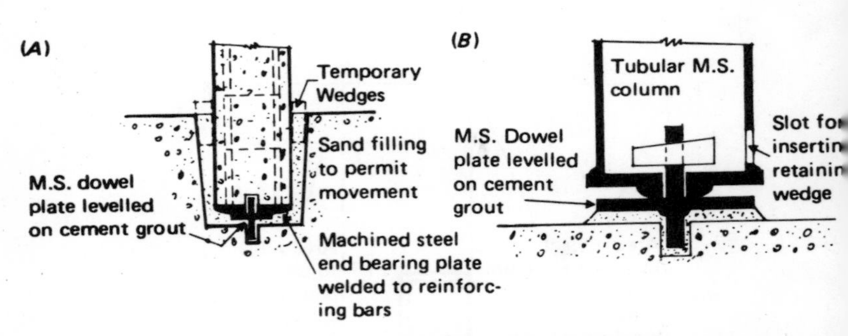

Fig. 44. Pin-jointed column bases for precast column and light tubular steel column.

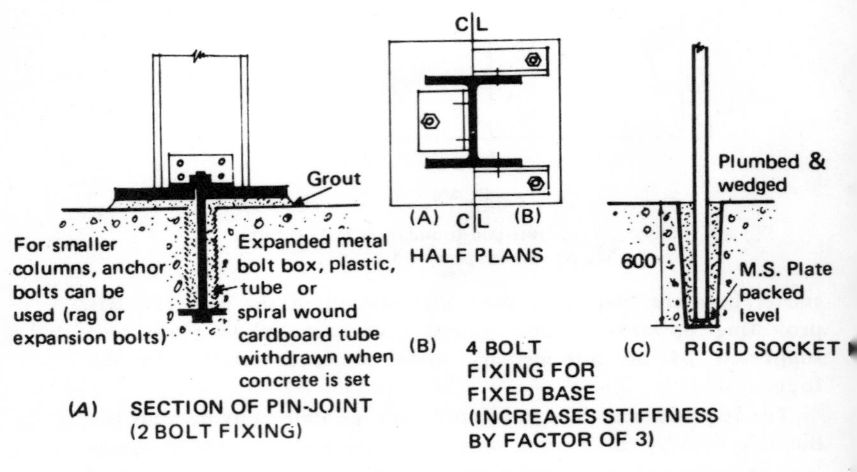

Fig. 45. Three types of end-fixing for steel columns.

require a pad of approximately 2 m^2 area. This is satisfactory until the situation arises where the pad foundations become almost continuous. Thus, columns at 6 m centres each carrying 450 kN on clay of only 50 kN/m^2 bearing capacity will require pads 3 m x 3 m and at or near this stage, the designer must turn to a raft foundation.

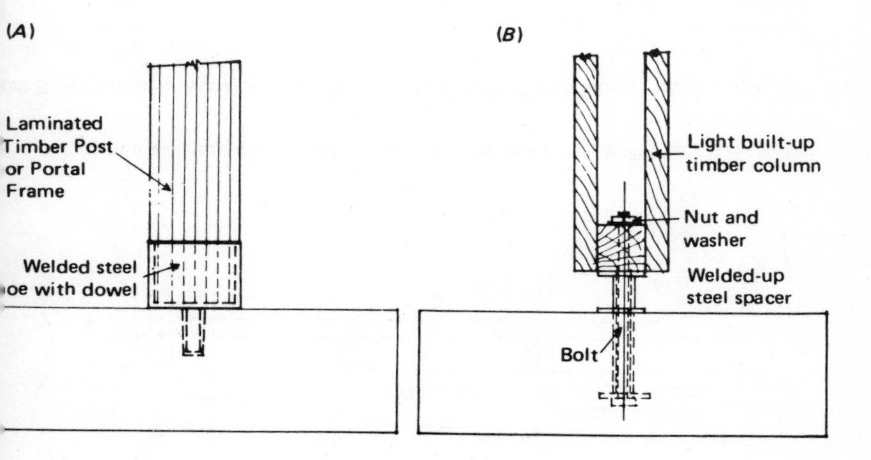

Fig. 46. Foundation shoes for timber posts.

Raft foundation

This consists of a continuous reinforced concrete slab under the whole building, taking all the downward loads and distributing them over an area large enough to avoid overstressing the soil beyond its bearing capacity. For a small house, a raft of 150–225 mm thickness of reinforced concrete (Fig. 47) may be sufficient; and for slightly larger buildings, rafts 225–375 mm in thickness may be used. For greater loadings, the raft can be stiffened by ground beams under walls, forming an egg-crate grid under the slab, and usually with an edge beam around the perimeter (Fig. 48). For tall buildings of considerable weight on good non-cohesive soil, a deep egg-crate sandwich raft can be used to withstand the load and spread it, and this may be of sufficient height to be of use for a basement storage area and for services installations with small openings connecting adjacent cells (Fig. 49).

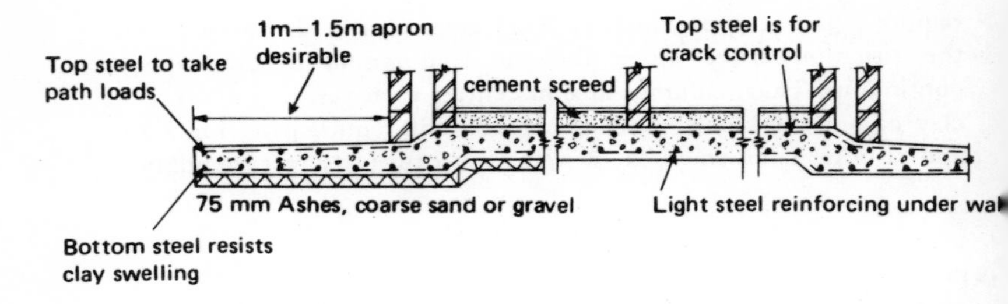

Fig. 47. Light reinforced concrete raft foundation to a house.

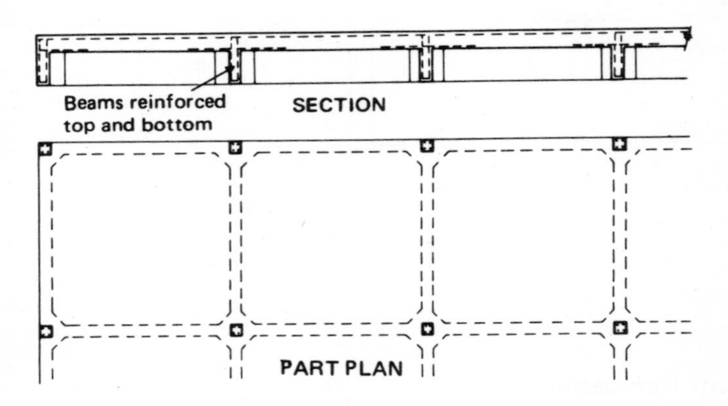

Fig. 48. "Waffle" raft foundation.

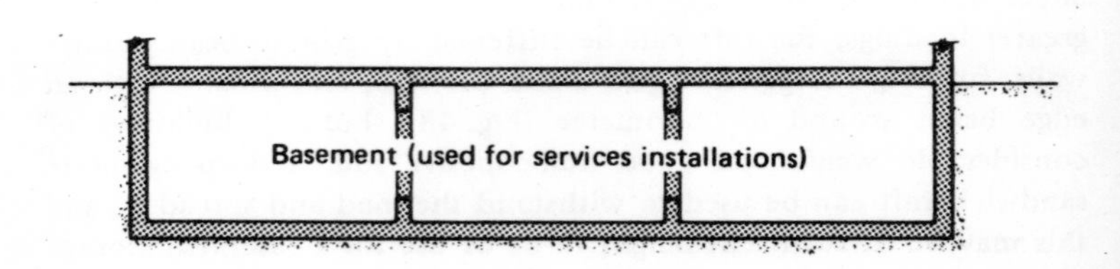

Fig. 49. Stiffened reinforced concrete basement acting as raft.

Buoyancy foundation or tanked basement

Where the bearing capacity of a soil is low near the surface, but the load of the building is so great that the site is not big enough to provide space for a large enough raft to spread the load, it is often possible to provide and utilize a large basement as a huge tank on which the building "floats" (Fig. 86). The weight of the building replaces the weight of soil excavated for the basement and the increase in weight on the soil under the bottom of the basement is often very small. One must remember that the deeper the soil, the more compressed it is, and the greater its bearing capacity. This can be a problem with clay, which begins to swell as soon as the weight of soil above it is removed. To obviate this, many contractors arrange marathon operations for pouring concrete for the basement slab as soon as the last of the clay is excavated. As an example, a contractor in S. California constructed a basement floor by pouring 3000 m^3 of concrete 900 mm thick in 10 h overnight, using five concrete pumps and 21 mixer trucks.

Deep basements, up to 24 m deep and containing several floors of car-parking, stores and service installations have been built to utilize this principle on the London clay. The London Hilton is an example, where the tower and basement weigh no more than the soil excavated on the site. It is important that such buildings should be designed symmetrically to avoid eccentric loading.

An example of excavation counterbalancing the weight of a building follows:

A site is available for a tall office block. Soil is soft clay of 110 kN/m^2 allowable bearing capacity and weighing 2000 kg/m^3. Assuming that a two-storey deep basement can be utilized 9 m deep, excavation to this depth will remove 18,000 kg/m^2 which equals approx. 180 kN/m^2. So soil at this depth will support 290 kN/m^2. Since modern office buildings weigh between 11,000 and 14,000 kg/m^2 of floor space, which is equivalent to a pressure of 11–14 kN/m^2, the building can be of 20 storeys, 18 above ground and 2 below,

i.e. 20×14 kN/m^2 (the higher figure given above)

$$= 280 \text{ kN/m}^2$$

or of 26 storeys, 24 above ground and two below if the lower figure of 11 kN/m^2 is taken. By increasing the depth of the basement, an even higher structure could be built.

Settlement of such a basement could quite easily be 75-100 mm and this would be considered reasonable. Up to 150 mm could be accepted if the basement substructure was carefully designed to minimize differential settlement.

Piled foundations

Foundations can be horizontal to spread the imposed load, or vertical (Figs 57a and 39) to carry the load down to strata more capable of providing the necessary support. This latter type of foundation can be a vertical concrete wall, but it is usually easier and more economical to concentrate the supports for the load on to the vertical columns sunk into the ground and known as piles. As this subject is complex, a separate chapter is devoted to it.

Chapter 7—Piles and piling

Piling is used:

1. When the load imposed by a building cannot be spread sufficiently over the available ground area without exceeding the allowable bearing capacity of the soil. This also includes the situation in which safe bearing capacity is not exceeded, but when settlement would be greater than is normally acceptable.
2. When settlement due to weak underlying strata is unpredictable but there is reliable rock or soil under which can be reached economically by modern methods. This includes building over mud and poor quality made-up ground.
3. When a building has to be founded on soils liable to shrinkage and swelling, usually seasonal.
4. When building over water.
5. When there are tensional forces tending to overturn or lift a structure. This can occur in several situations such as:
   (a) With side wind force on a high building.
   (b) When tension members such as the cables of large tent-like or "tensegrity" structures are anchored down to the ground.
   (c) When there is need to resist uplift from ground swelling after the removal of the weight of many tons of soil at the bottom of a deep excavation, or on absorbing moisture.
6. To resist lateral loads. E.g. the piles can be driven angled to resist the angled resultant force on a foundation of a retaining wall, or the side impacts from ships alongside a jetty.
7. To underpin or strengthen existing foundations (Fig. 50).

From the above, it can be seen that it is always important to consider the direction of the forces which will act upon the pile and to ensure that the pile is designed in the best way to resist them, whether they be compressional or tensional, vertical or angled. The other consideration is to make certain that the resistance of the soil is sufficiently utilized to withstand the forces transmitted by the pile. In all but the weakest soils, friction between circumference of pile and soil is very important. From this point of view, piles with rough surface texture formed in situ by contact with the sides of the borehole can be of advantage, but a smooth surface pile is easier to drive.

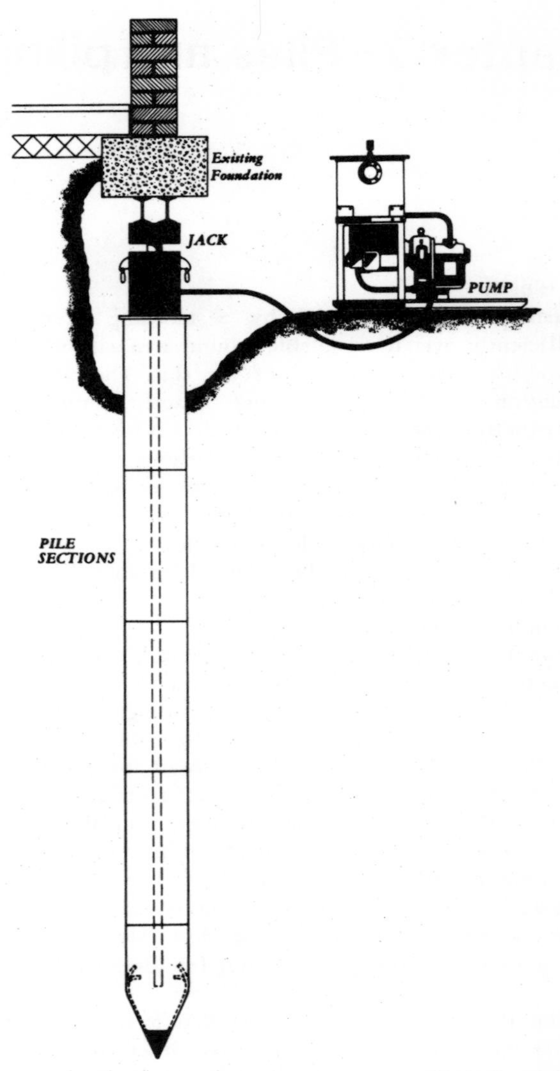

Fig. 50. Jacked-in sectional precast concrete piles used to strengthen existing foundations.

The effectiveness of a pile may depend on:

(a) friction and adhesion to soil in contact with the perimeter;

(b) end bearing only (in compression usually, but tension piles may use enlarged ends for bearing on the soil above);

(c) a combination of (a) and (b)

and must be designed or selected accordingly. The material used, type of pile, method of driving or forming, must be selected for the individual circumstances of each job. Fortunately, there are many alternatives from which to choose, as engineers and specialist firms have evolved systems, mostly patented, to suit every situation, and there is much commercial competition in this field.

Materials used for piles

Timber These have been used since prehistoric times, when lake-dwellers built their huts on piles of tree trunks driven vertically into the mud. They are often perfectly preserved when found today in the acid waters of peat bogs. Generally, a suitable timber will last a long time if kept either in permanently wet or dry conditions, but not if these conditions alternate. It is wise to impregnate timber piles with preservative under pressure before driving, against fungus attack and insects. Marine borers are a danger to timber piles in sea water. Pine, fir and larch are mostly used for cheapness, but suitable hardwoods are elm, oak, teak, hornbeam and greenheart. The last three are expensive but sometimes essential in marine situations to resist impact, etc. The stripped natural log is cheapest, but squared sections are best where bracing is to be attached. The piles can be of considerable length in one log, or spliced into even greater lengths, but precautions are needed in driving. Tops are usually steel capped or banded in hoop iron to prevent splitting and the points protected by steel points and straps. Driving should be by light blows with butt end up, and working load should not exceed 25 tf for even the largest pile. Such piling today is mostly used in "out-back" areas where timber is the cheapest or only material available, or in building over water (Fig. 51).

Steel Individual piles are of H section or hollow tube or box section filled with concrete after driving. Sheet steel piling, consisting of a number of corrugated profile interlocking hot rolled sections is widely used, and has many applications for earth or water-retaining foundation walls and basements (Fig. 52). The steel may have copper content to inhibit corrosion or be painted with bituminous paint for protection. Steel piles can be driven hard, withdrawn if required and lengthened or cut fairly easily.

Concrete These piles can be sub-divided into:

(a) Pre-cast.
(b) In situ.

(a) *Pre-cast piles* These can be sub-divided again into:

1. Sheet piles (Fig. 53).
2. Column piles (Fig. 54).

and they can be constructed in ordinary reinforced concrete, pre-stressed concrete, or in fibre reinforced concrete (using glass, plastic or steel wire fibres). Conventional reinforcement or pre-

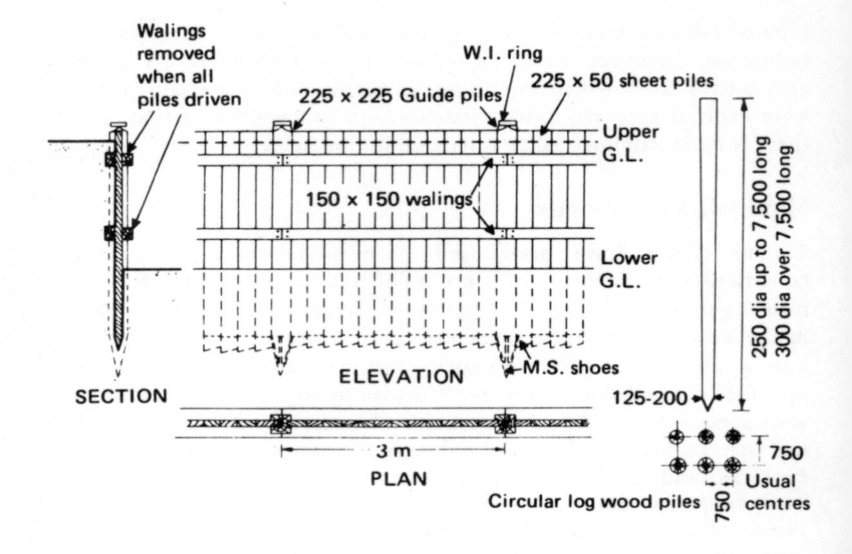

Fig. 51. Timber piling for retention and support.

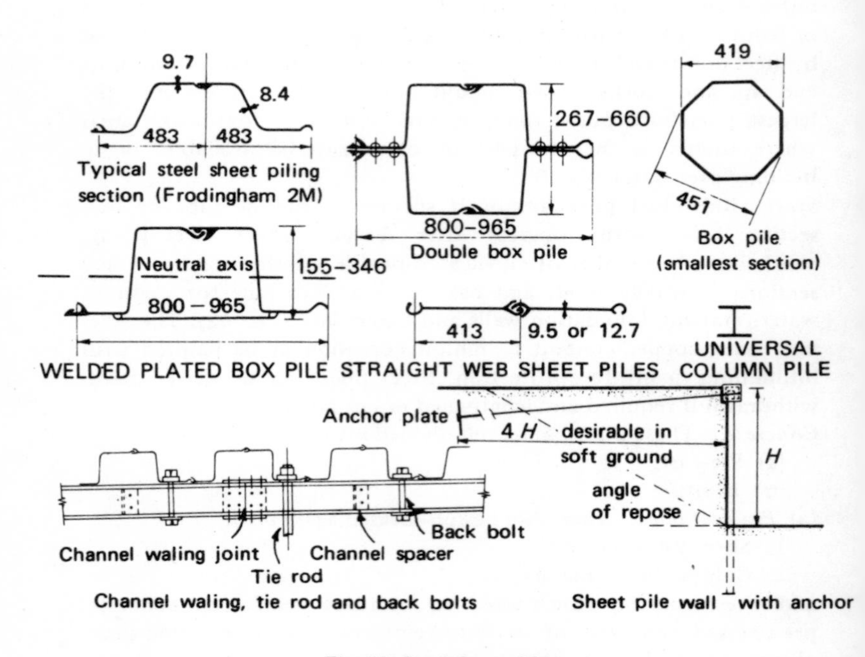

Fig. 52. Steel sheet piling.

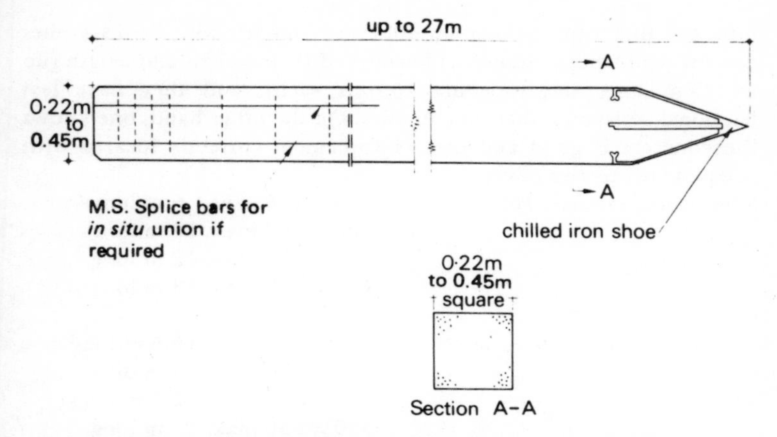

Fig. 53. General details of "Bison" pre-stressed concrete bearing piles.

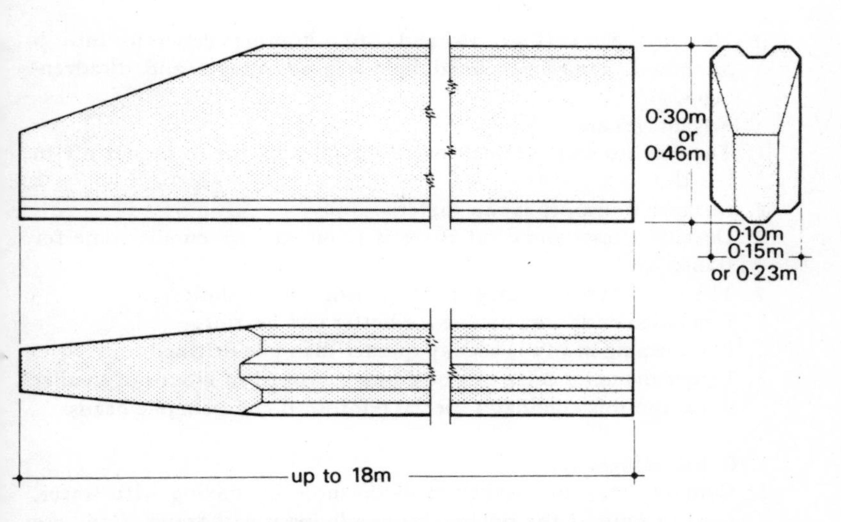

Fig. 54. Detail of "Bison" pre-stressed concrete sheet pile.

stressing is used for compression piles, pre-stressed concrete for tension piles, sheet piles, and for piles driven in open water. It is easy to break away concrete at the top of a driven conventionally reinforced pile to expose enough steel reinforcement to use as starter bars for pile head reinforcement, but in pre-stressed piles this is not possible and special starter bars have to be provided in the top of the pile when casting, or the head of the pile may penetrate 75-300 mm into the pile cap, relying on adhesion and friction. Pre-cast piles present problems of transport because of their weight and length (up to 27 m long and weighing 15 tons each), and they have less frictional resistance than in situ ones. On the other hand, one knows the concrete is good and the reinforcement correctly located with adequate protective cover.

Sheet piling sections are:

300 × 100 max. 6 m long
300 × 150 max. 12 m long
450 × 150 max. 12 m long
450 × 225 max. 18 m long

Pre-stressed pile sections are:

225 × 225 max. 10.5 m long
250 × 250 max. 15 m long
300 × 300 max. 22.5 m long
350 × 350 max. 27 m long
400 × 400 max. 27 m long
450 × 450 max. 27 m long

(b) *In situ piles* (Figs 55 and 56) Pouring concrete into a pre-bored, sometimes lined hole has advantages and disadvantages.

Advantages are:

1. There is no wastage with over-long piles having to be cut off to length.
2. Cheaper transport of raw materials only.
3. Usually a less complicated rig is required, less cumbersome for transport.
4. There is better friction grip on the sides of boreholes.
5. Expanded ends are possible for better end bearing.
6. Pile forming in situ is usually quieter than pile-driving.
7. Larger diameter piles can be formed instead of groups of smaller piles, and this eliminates special reinforced concrete pile heads.

Disadvantages are:

1. Concrete may be weakened in the hole by mixing with water, sand or mud at the bottom, or as a lining is withdrawn. Concrete should not be of less strength than standard mix C.P. 114-21 N/mm^2.
2. Reinforcement may be displaced or given insufficient cover during operations on site.
3. Driving of new piles alongside freshly poured ones may damage them before they have hardened.
4. They cannot easily be withdrawn or re-driven if required (not often required).
5. They are not as flexible as pre-stressed piles (and in some situations flexibility is desirable).

6. The light linings and boring tools employed may be damaged by hard obstacles such as boulders.
7. In situ piles are unsuitable for very soft soils unless a lining is used.

Maximum bearing capacity of concrete column piles

As a guide to the size and number of piles required in a foundation, the following information can be useful providing it is realized that every job must be assessed carefully and individually in practice.

Maximum bearing capacities of concrete column piles.

Circular piles

Dia. in mm	325-350	350-375	400-450	450-500	500-550	600
Max. load in tonnes	35	50	70	90	110	140
Max. length	10 m	13 m	16–18 m	18-21 m	21-24 m	27-30 m

Large diameter piles (concrete stress 5.3 N/mm^2) (1 : 2 : 4 mix)

Dia. in mm	750	900	1050	1200	1350	1500	1650	1800	1950	2100
Max. load in tonnes	237	341	464	606	767	947	1146	1363	1600	1857

Soil conditions affecting choice of pile type

Type of ground and soil conditions are often the determining factors in choosing the tvpe of pile to use for a particular job. The following notes are intended as a general guide:

1. *Made-up or very soft ground overlying hard strata* Use end-bearing piles, possibly with widened bases. Such piles can be up to 30 m long.
2. *Medium strength subsoil of great depth* Use friction piles obtaining their support by friction and possibly adhesion between subsoil and surface of pile plus some end-bearing which can be increased if desired by enlarging the bases.
3. *In water or saturated soils, and in non-cohesive soils* Use driven pre-cast piles (the vibration consolidates dry sand).
4. *Clays* In situ piles are usually best, as driving pre-cast piles tends to destroy friction and adhesion in stiff to hard clay, due to whip in the shaft forming a gap around it.

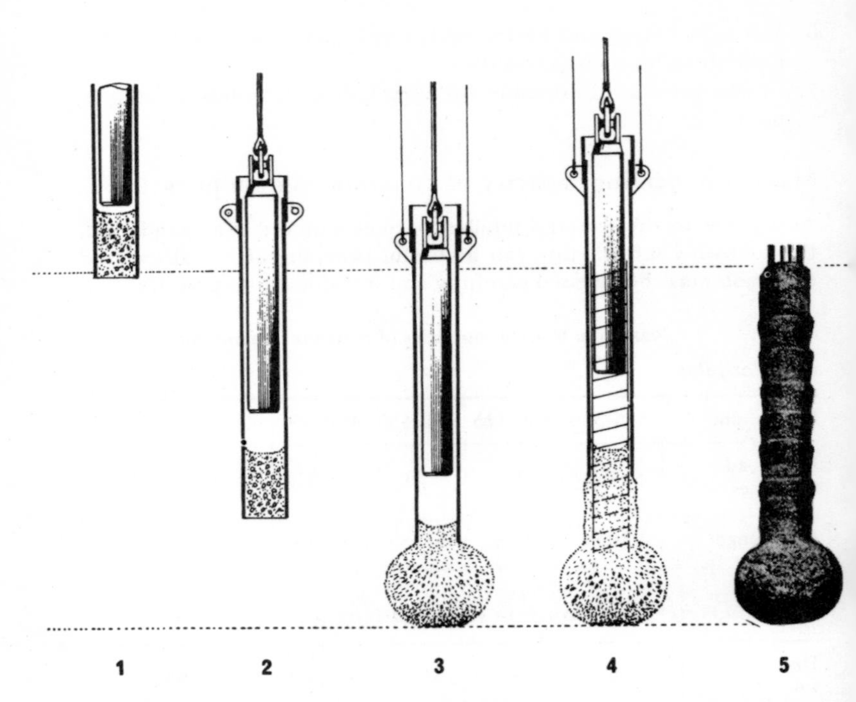

Fig. 55. Formation of in situ pile with bulb end. The diagrams show the complete driving sequence.

1. Consolidating the 600-900 mm of aggregate to form a solid plug
2. Driving the tube
3. Forming the base
4. Forming the shaft
5. Completed pile.

5. *Soils liable to expansion and shrinking* Use short bored piles to anchor lightly loaded buildings by friction and end-bearing to soil below the level of seasonal movement (Fig. 57). See also B.R.S. Digests Nos. 3 and 42 (1st series).
6. *Fine sand* Confine by surrounding building area with sheet piling, to prevent erosion by wind.

Methods of pile-driving, and of forming in situ piles

The methods listed below will enable a designer to appreciate the varying methods available for sinking piles. No attempt has been made to cover all available systems, but those of particular value in special cirumstances have been mentioned.

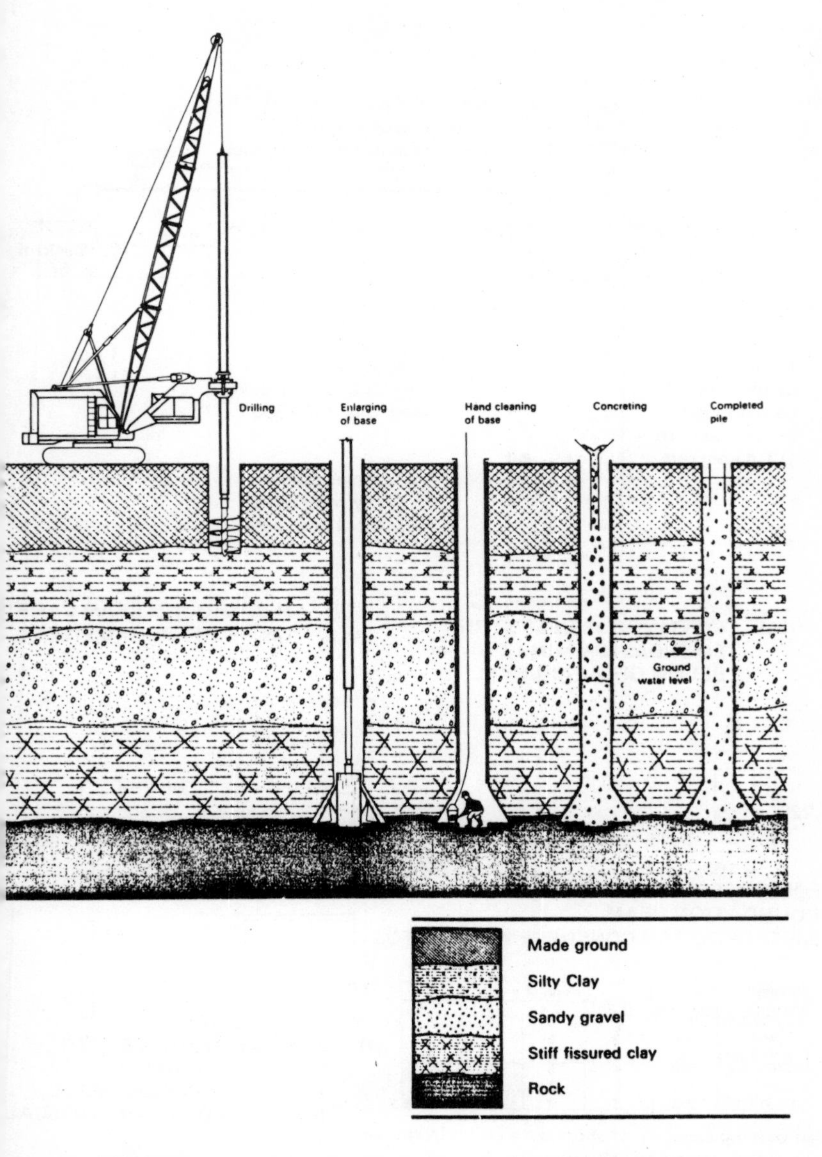

Fig. 56. Stages in construction of large-diameter augered foundations with reamed-out bases.

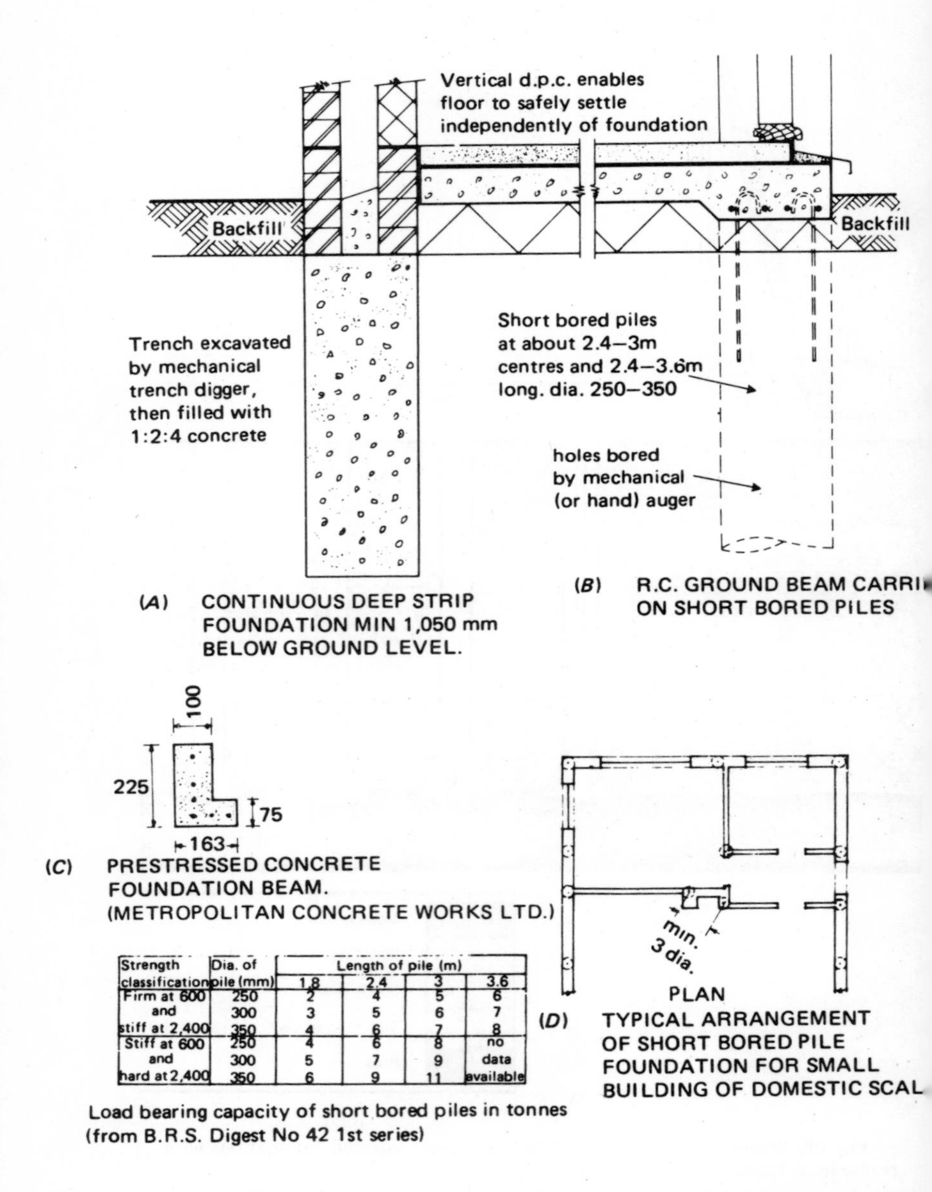

Strength classification	Dia. of pile (mm)	Length of pile (m)			
		1.8	2.4	3	3.6
Firm at 600 and stiff at 2,400	250	2	4	5	6
	300	3	5	6	7
	350	4	6	7	8
Stiff at 600 and hard at 2,400	250	4	6	8	no data available
	300	5	7	9	
	350	6	9	11	

Load bearing capacity of short bored piles in tonnes
(from B.R.S. Digest No 42 1st series)

Fig. 57. Foundations suitable for light loading on firm to hard clay subsoils.

1. *Single-acting pile hammer* This is the simplest pile-driving rig and the one which probably gives rise to most complaints of noise as the hammer is raised and dropped on the head of the pile for a controlled distance, accompanied by the sound of escaping steam or the thunderous roar of an air compressor. It is still commonly used for driving pre-cast concrete, timber and sheet piles.
2. *Double-acting pile hammers* These keep up a rapid succession of blows on the pile, thus driving faster and vibrating the ground at the same time, which is helpful with sands and gravels.
3. *Vibratory pile drivers* These vibrate the pile to temporarily negate friction between pile and soil so that it sinks in rapidly (200 mm/s) and quietly to the required depth. Vibrations are dampened down by the soil within 1 m of the pile, so neighbouring buildings are unaffected. There are several vibratory systems available on the market, among them being the excellent Bodine Resonant Pile Driver (G.K.N. Foundations Ltd.) which will drive both sheet piles and pre-cast concrete piles as long and as deep as is required. This driver works at very fast speeds in both non-cohesive and cohesive soils.
4. *Hydraulic ram pile drivers* These are very silent, and impart no vibration to the ground, so are most useful in avoiding both annoyance to people and damage to existing buildings. They are therefore considerably employed for pile-driving to underpin failing foundations and can even be used inside a building with low head-room. This is made possible by the use of short sections of pre-cast piles jacked in with dowelled joints. Sectional piles can also be post-tensioned. Resistance for the jack can be obtained by jacking against the existing foundation if underpinning (Fig. 50), or against a heavy kentledge. A system known as the Taywood Silent Pilemaster (Fig. 58) is available for jacking-in sheet piles. This is a heavy machine consisting of a power pack and a battery of eight hydraulic rams each with a 225 t/750 mm stroke. It is lifted on to a group of eight sheet piles by a crane which holds it in position until the piles become stable after being jacked in two at a time against the reaction of the other piles. After all jacks have been fully extended by jacking in the eight piles for 750 mm, the machine lowers itself by retracting all rams and recommences the operation. It can drive 20 piles 9 m long into the London clay in one day.
5. *Screw-pile systems* Pointed piles of concrete or steel are fitted with helical steel blades and are screwed into the soil by applying torque to the head of the pile or to a mandrel fitted inside a hollow cylindrical pile. They have good bearing capacity, but cannot penetrate very deep into firm or dense soils.

Fig. 58. "Silent" hydraulic ram sheet-pile driver ("Taywood Pilemaster").

6. *Driven hollow pile or lining systems* There are a number of slightly varying systems whereby a light hollow concrete pile or steel lining tube is sunk in sections while the centre core is excavated until the final depth is reached (Fig. 59). Some types are driven by a hammer falling on a concrete nose cap at the bottom of the hollow lining, and this is obviously less noisy than the surface pile hammer method. A reinforcing cage can be lowered down the shaft and several alternative operations carried out, e.g.:
   (a) packing the interior of the tube with concrete when it has been driven;
   (b) packing the shaft with concrete as the lining is withdrawn;
   (c) withdrawing the lining a short way and then forcing concrete out at the bottom of the shaft to form a "bulb-end" pile before withdrawing the rest of the lining and filling the shaft (Fig. 55). Some systems use an airtight cap and compressed air to raise the lining pneumatically.
7. *Auger bored holes for piles* Vertical rigs with large diameter augers (Fig. 60) are now available for boring into cohesive soils. These can work inside a lining if necessary but are normally used where the clays are stiff enough not to need support. Holes can be bored up to 2.4 m dia. and over 30 m deep. Bases can be underreamed for bulb or splayed feet up to 5.4 m dia. Inspection can be carried out before lowering reinforcement and filling with concrete. Except at the top, no reinforcement is required for piles over 900 mm dia. It is due to this system that the large diameter bored pile foundation has become an important cheaper and quicker solution than driving multiple piles and forming pile heads where large loads have to be carried on clay (Fig. 56). Upwards of 2000 t have been carried on one pile. They are particularly suitable for London clay, which with increasing depth, increases in strength and decreases in compressibility. They are nearly always treated as end-bearing piles when bases are underreamed. "Belled bases" (as they are known when splayed at about 55°) are enlarged up to a maximum of 2½-3 times the shaft diameter and must be at least 300 mm clear distance apart from neighbouring bases.

   Short bored piles of small diameter for house foundations can be bored out with hand augers in suitable clay, but mechanical augers of a portable nature are also available for this job on small sites.

Test piles

Whenever possible, test piles should be driven on a site and tested, preferably to the ultimate load. This is usually taken to be reached either when the pile continues to settle without further increase of

A

B

Fig. 59. Bored pile rig, and operational sequence.
A. Bored piling rig operating in limited headroom
B. Commencement of boring and sinking tube
C. Use of tube lifter for placing or removal of sections of tube.
D. Hammer compacting concrete base
E. Placing cage of reinforcement after completion of base
F. Jacking out tubes whilst concreting pile

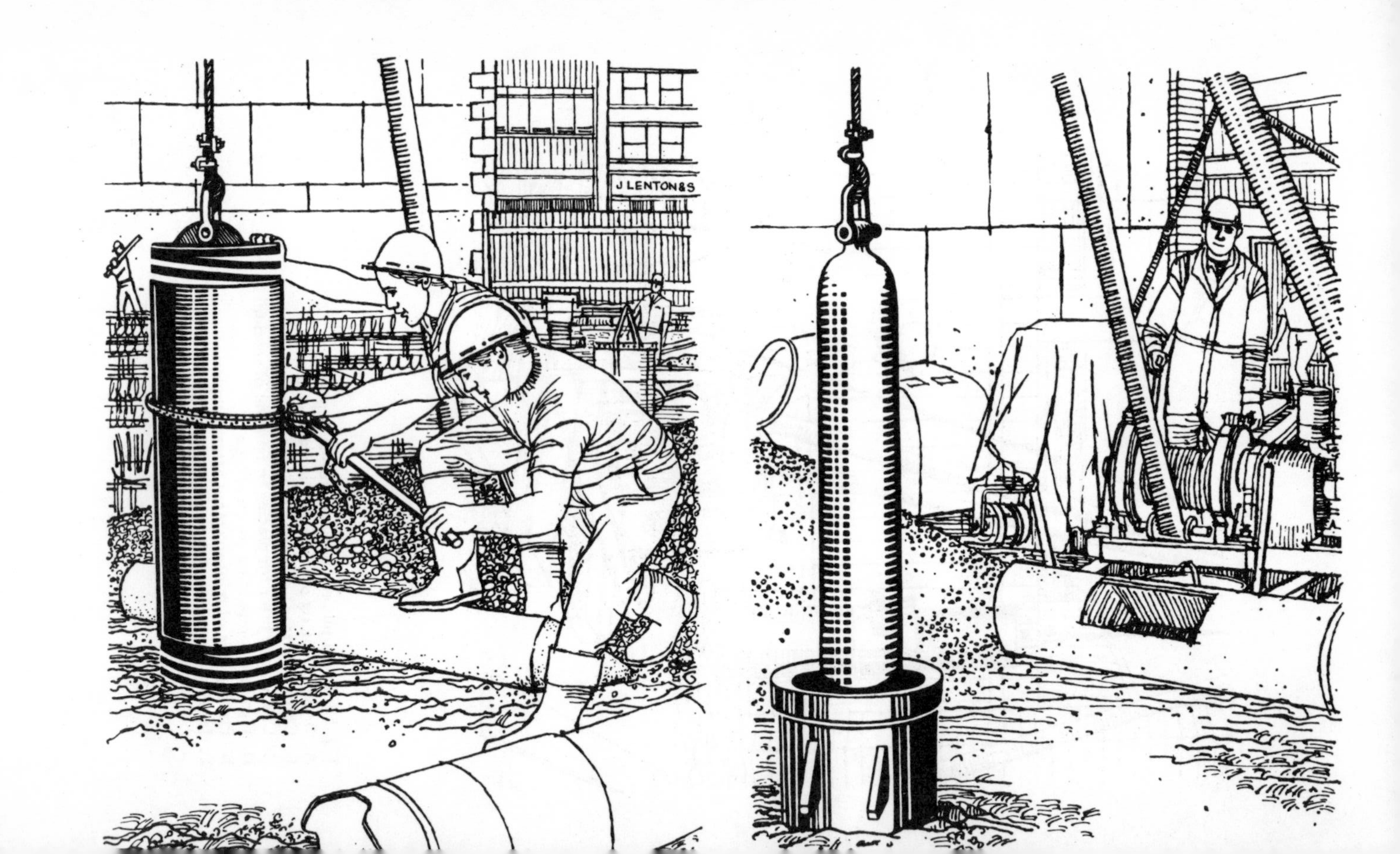
J LENTON&S

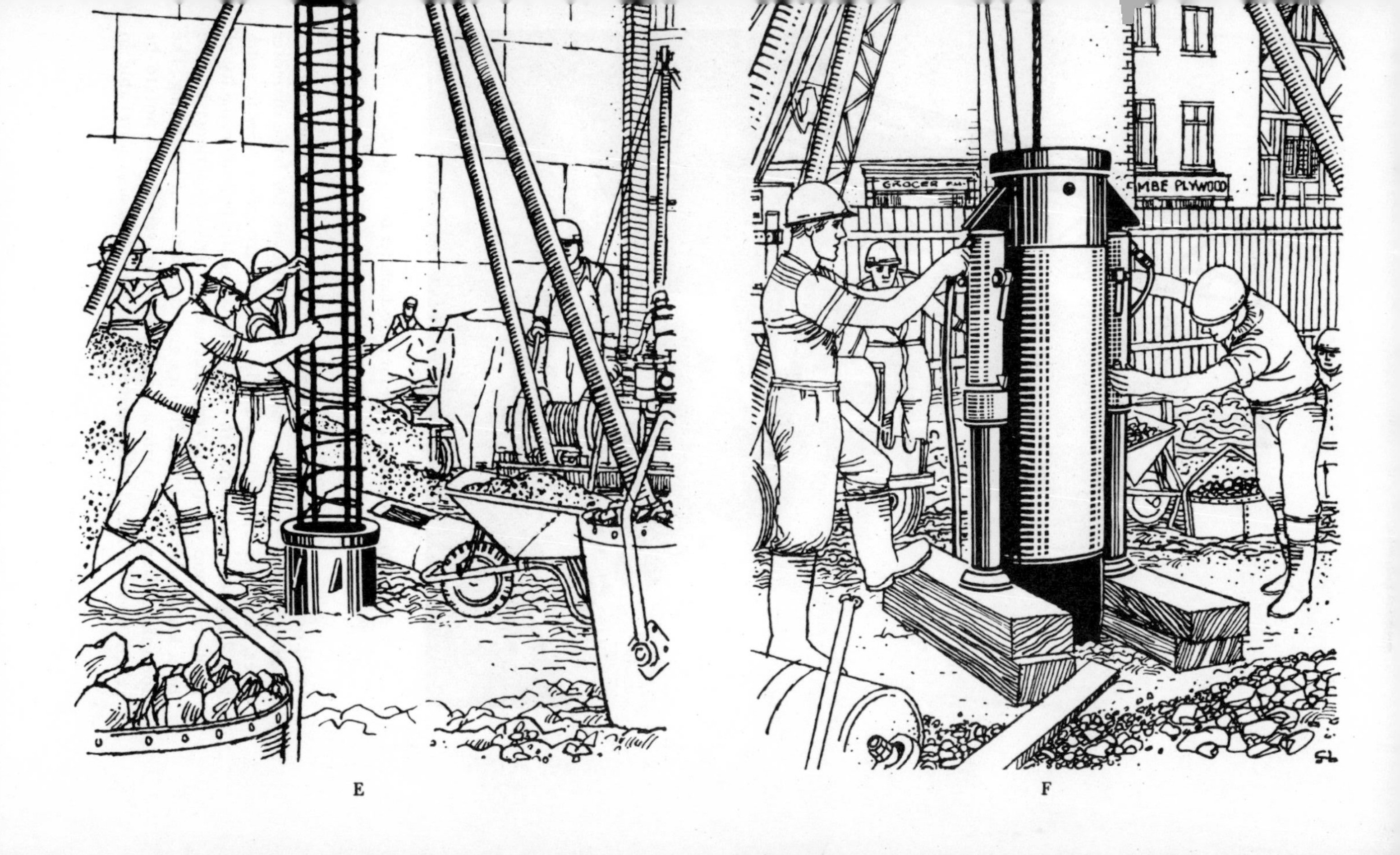

E

F

Fig. 60. Mechanical auger for large-diameter bored piles.

load, or when the settlement reaches one-tenth of the pile diameter or least width. This is not acceptable, however, for large diameter piles. A factor of safety of 2-3 is usual for groups of piles in sand, and 2-$2\frac{1}{2}$ for a single pile in clay. Estimating the bearing capacity of a pile or group of piles is very much a matter of experience backed by a knowledge of soil mechanics and the results of test piling. The formulae offered by various authorities are varied and not to be relied upon alone. Unfortunately, test piling is also unreliable in many cases, especially with bored in situ piles. As may be imagined these can vary enormously in bearing capacity. One pile cannot be test-loaded to represent a large diameter pile carrying a 1000 t, neither can it truly represent the bearing capacity of a group of piles. Ten piles in a group will probably carry more than 10 times the load on a single test pile if in sand, but less if in clay. The bulb of pressure of a group of piles may result in overloading of a deep layer of soft clay, yet this would not be affected by the pressure from a single pile (Fig. 24). Rotational failure of a whole group of piles acting as one is possible in cohesive soils (Fig. 13), and block settlement can occur on failure in non-cohesive soils.

The method of test-loading is similar to that given for a plate loading test, but sometimes a pile to be tested is jacked down against the uplift resistance of 2-4 surrounding piles (Fig. 61). If it is possible to arrange this saves on kentledge (Fig. 62). Jacked-in piles are tested as they are driven in.

There is also the Constant Rate of Penetration Test (C.R.P. test) devised by Dr. Whitaker of the B.R.S., but so far mostly used for research.

Design of pile groups and pile caps

Loads on foundations are often very great and the soil may be unsuitable for large diameter single piles. In such circumstances groups of piles can be driven and united into a single base by means of a reinforced concrete pile head.

Grouping of piles is fairly standardized and details of types of pile caps for from 2-19 piles can be found in specialized publications of reinforced concrete detailing. Examples of pile caps are shown in Fig. 63. Owing to the difficulty of driving piles with extreme accuracy, it is often inadvisable to use one or two piles only for a base as there may be considerable eccentricity when a column is supported on them. A group of three or more columns allows greater tolerance. Spacing of friction piles in groups should be at 2½-3 times the diameter of the piles, but not less than 1050 mm or the circumference or least dimension of the pile (ref. C.E.C.P. 4 Para. 3.22). Spacing of end-bearing piles can be reduced to 750 mm, or two diameters, whichever is greater.

Pile caps may be anything from 550 mm to 1500 mm thick, dependent on the load to be distributed and the number of piles and spacing. Where there are only one or two piles, the pile caps must be linked by r.c. tie beams to prevent lateral movement, but the beams are often needed anyway to support floor slabs or walls over, transferring their loads to the piles instead of to the soft ground between.

There are occasions when the engineer will call for the testing of a pile on a particular site. The pile to be tested is usually chosen at random.

The most frequently used method of testing is by jacking against anchor piles. To do this special anchor piles are installed on either side of the pile to be tested, and these two are connected by R.S.J's. The load is then applied to the pile by means of a hydraulic ram jack between the head of the test pile and the R.S.J.'s. Pressure is gradually applied and any subsequent deflection of the pile head is recorded by means of dial gauges.

Piles are normally tested to 50 per cent beyond the stipulated working load.

Fig. 61. Testing a pile, by jacking against rolled steel joists connecting four anchor piles.

Fig. 62. Test loading a pile, using 500 tonnes of kentledge.

Problems in piling

A number of problems peculiar to piling frequently occur. These are:

Over consolidation with sand When a group of piles is driven into sand, the soil becomes more and more consolidated and it could be difficult to drive the central piles. It is better either to work from the centre outwards or from one end of the group.

Ground heave in clay This tends to occur as a group of piles is driven, and will lift central piles already driven. With some methods of driving, it may be difficult to re-drive the piles to the correct level. Also the ground tends to consolidate under load after a time, and piles in the centre of the group are dragged down with the soil.

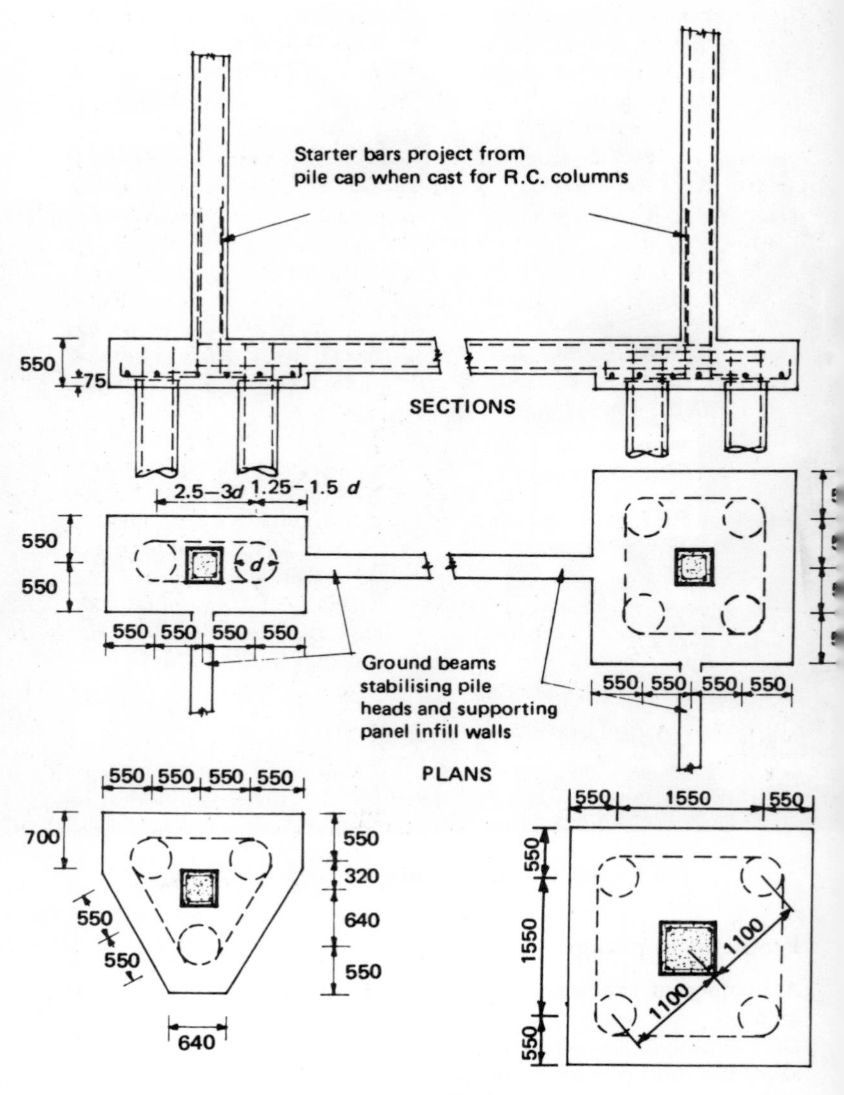

Fig. 63. Pile caps.

This is why bored in situ piles are best on clay soils.

Boulders in the soil Many piles are broken or deflected by boulders, but most piling contractors have various means of splitting, drilling through or exploding the obstacle. In very difficult stony conditions, the steel H pile is very suitable to use.

Rock under very soft ground In this case it may be insufficient for piles merely to rest on the bedrock. Holes should be drilled down into the rock, and hollow piles used which can have reinforcement lowered through them into the rock hole before filling with concrete. Post tensioning to the rock can also be used.

Piles subject to vibration Since piles can be sunk by vibratory means, there is obviously a danger if, for instance, vibrating machinery is installed above a piled foundation. The biggest danger is with sand which may consolidate and pull down the piles.

Cavities in strata When a group of piles is driven, it is possible for the top strata penetrated to bridge between piles. This means that if and when the lower strata consolidates, it may shrink downwards, leaving cavities between top and lower layers, thus reducing frictional area of the pile and soil acting together. This actually happened on a site in the London docks.

An awful example of cavities occurring under thin top strata can be quoted from Southsea, Hants, where many small buildings near the South Parade Pier are founded on short piles, adequate for the small loads carried. A contractor attempting to drive longer piles for a much larger building, found his first pile disappeared into the ground and was never seen again. Investigations disclosed the presence of a very extensive cavity containing an underground lake. Today, this would have been discovered much earlier by a soil investigation borehole, a rare occurrence at that time.

Economics of piling

Piling has generally been considered expensive, but this is not always so. It has been proved that on London clay, it is cheaper to use short bored piles than the deep strip foundations otherwise required for this type of ground. Machinery is now readily available for augering this clay for larger, longer piles. When compared with the cost of building a basement; excavation, shoring, pumping, waterproofing and the time involved, piling may well prove cheaper if there is a choice.

When checking tenders for piling from invited sub-contractors, compare the length of piles on which the quotation is based, and the cost per metre for any extra length required.

Caissons

Used in large-scale building or civil engineering works, caissons are of three main types:

(a) Open caissons.
(b) Closed caissons.
(c) Pneumatic caissons.

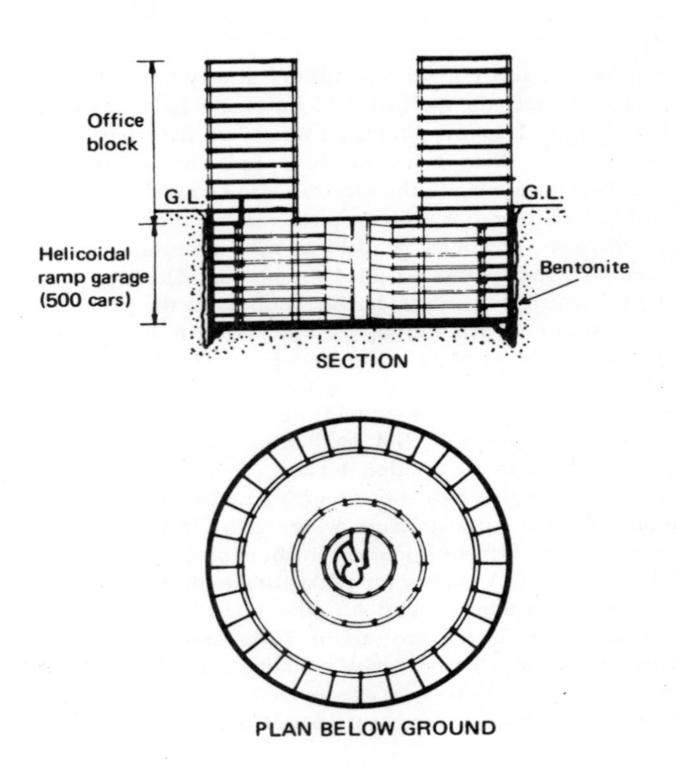

Fig. 64. Office block in Geneva: Seven-storey helicoidal ramp underground garage, constructed by sinking 57 m diameter caisson. (Bentonite used as a lubricant.)

Open caissons

Open caissons are box or cylinder type structures open at top and bottom which can be sunk into the ground by excavating or water-jetting soil from under the lower chisel-shaped cutting edges of the vertical walls. Large caissons will usually sink under their own weight, but skin friction with the soil can be reduced with the aid of Bentonite around the perimeter (Lorenz-Fehlmann technique, Fig. 64). They are being used more and more as an economical way of constructing basements, underground car parks, etc. Smaller caissons have been used as foundations in a similar way to large diameter bored piles. The John Hancock Centre at Chicago, founded on 2.4 m dia. caissons and 100 storeys high showed signs of excessive settlement while still unfinished, but this was found to be due to defective concrete in the caissons; clay, earth and sand from the soil sides mixing in with the concrete when it was being poured. This is a danger with all in situ foundations.

Closed caissons
Even when built in thick reinforced concrete, closed caissons will float like a ship. They can therefore be used for foundations in the sea, by floating them into position, and then opening valves to let in sea water, or loading them until they rest on the sea bed. This method was used to build the Nab Tower in the First World War, the Mulberry Harbour in the Second World War and more recently, a lighthouse on the Goodwin Sands.

Pneumatic caissons
These are referred to later (Chapter 10) as temporary works for excavation under water, but they can be equally well used for permanent caissons sunk in ground where there is a heavy influx of water.

Further references on piles and piling

Piling Techniques — Building and Contract Journals Ltd.
B.R.S. Digest (2nd series) No. 95 Choosing a type of pile — H.M.S.O.
Proceedings of the Symposium on Large Bored Piles 1966 — I.C.E.
Literature from pile manufacturers and piling contractors

Chapter 8—Retaining walls

Such walls may have to retain:

Solids Usually natural soil undisturbed except next to the wall itself, but sometimes fill of varying materials when a site is terraced. Other materials such as solid fuel, chemicals, etc. may need to be stored within retaining walls on a site.

Liquids Sea, river and reservoir walls are examples and occasionally there may be need to contain liquids other than water, such as effluents from factories, oil, and even liquid mud or clay.

A combination of solids and liquid acting independently, as with gravel soil with a high ground water level, or where there is water on one side and soil on the other.

Solids may well be banked up higher than the top of the wall. In this case, the wall is described as a *surcharged retaining wall.*

There may well be other loads on the wall, such as dead loads from a superimposed building or live loads, as referred to later. Essentially however, a retaining wall is one which resists lateral loads.

The design of retaining walls can be very complex, as many factors are involved. Failure may occur through:

(a) *Forward sliding* (Fig. 65a).
(b) *Overturning* (Fig. 65b).
(c) *Settlement* due to overloading the soil under (often eccentric loading causes tilting forward) (Fig. 65b).
(d) *Bulging or bending* (with flexible walls) (Fig. 66).
(e) *Circular slip plane movement* of the soil behind and under (Figs 65c and d).
(f) *Inadequate anchorage of sheet piling* (Fig. 65d).

Design information required

Primary information required in connection with the design of retaining walls is as follows:

1. *Situation.*
2. *Height required.*
3. *What is to be retained?*

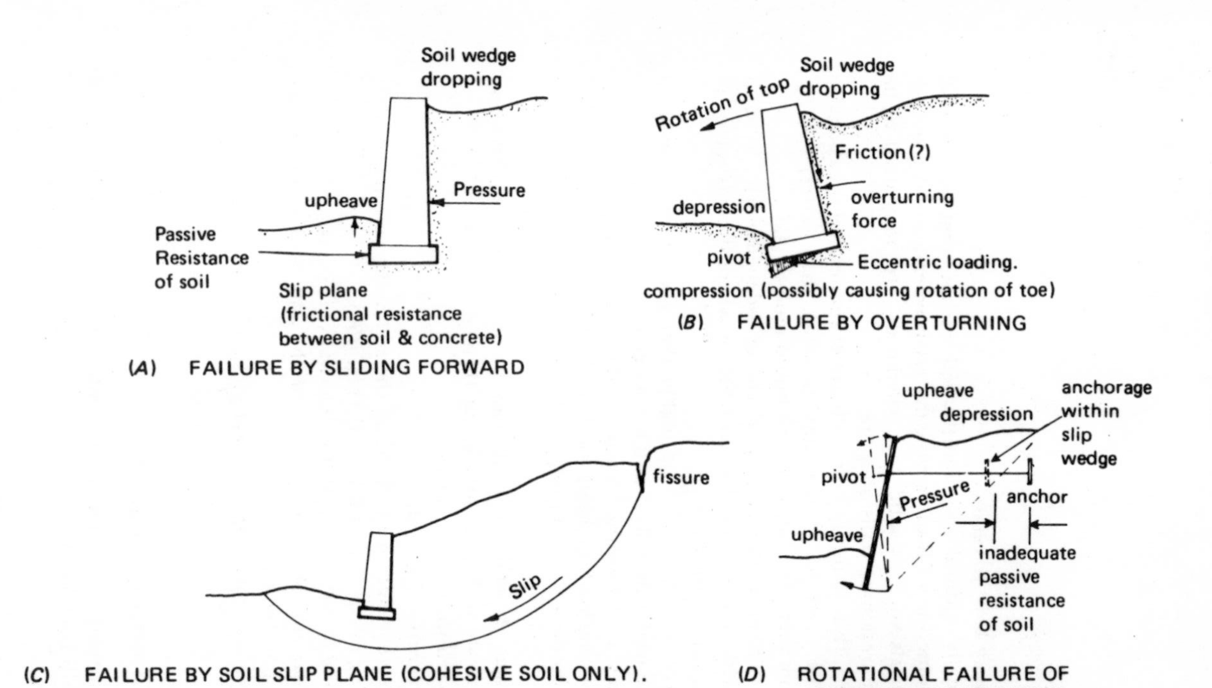

Fig. 65. Some causes of failure of retaining walls.

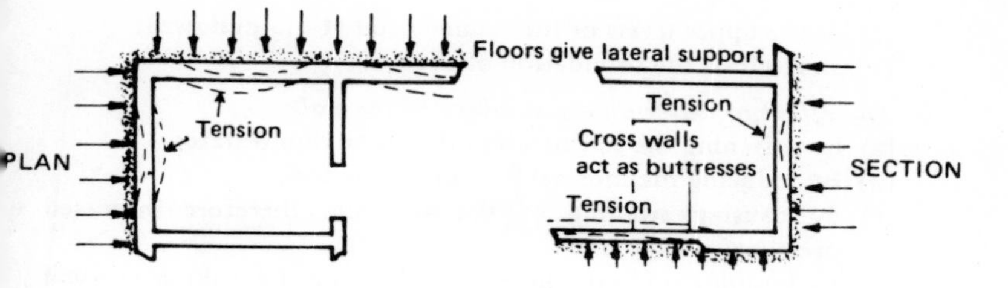

Fig. 66. Basement walls acting as retaining walls.

4. *To what height?*
   (a) level with top of wall;
   (b) below level of top of wall;
   (c) in the case of solids, possibly surcharged and if so, at what angle? (Usually the natural angle of repose, equal to the internal angle of friction.)
5. *Geology and nature of soil or rock*
   (a) behind the wall;
   (b) on which the wall will be founded.

Different methods are used to calculate earth pressures on a wall dependent on whether the soil is

(a) *non-cohesive:*	dry
	moist
	saturated
(b) *cohesive:*	non-fissured clay
	silts or partially saturated clay
	stiff fissured clay

and it is obviously necessary to know the angle of internal friction of the soil, as only above this angle will the soil be likely to slide and press against the wall.

For the soil under the wall, it is necessary to know:

(a) the nature of the geological strata beneath for a depth at least equal to twice the breadth of the foundation;
(b) in cohesive soils whether there is any possible slip plane;
(c) the ultimate strength;
(d) compressibility;
(e) moisture content;
(f) liquid limit;
(g) coefficient of friction between soil and base of foundation;
(h) sulphate content.

Generally, the effect of frost and long-term weathering of the soil, which may alter its characteristics and strength, is likely to be:

(a) at the upper levels of the retained soil (1-1.5 m down);
(b) in front of the foundation of the wall.

6. *The effect of water on the stability of the wall:*
   (a) by softening the soil to a liquid or semi-liquid state;
   (b) by reducing the internal friction of the soil;
   (c) by causing swelling of the soil and therefore increased pressure;
   (d) by building up hydraulic pressure behind the wall; or in some cases, in front. Variation of head of water, as in tidal or seasonal changes of level, must be anticipated;
   (e) by lubricating the friction surface under the foundation;
   (f) by lubricating a possible slip plane in cohesive soils and starting a landslip involving the wall and retained soil (Fig. 65c);
   (g) by scouring and erosion under foundations;
   (h) by impact of waves.

7. *Additional dead or live loading on the wall* such as:
   (a) the weight of building resting on the retaining wall of a basement;
   (b) The load of a foundation of a building on soil near the top of the wall (increasing lateral load and possibly the likelihood of a landslip on cohesive soil);
   (c) earthquake forces (lateral) in some parts of the world;
   (d) moving loads, such as cranes and heavy lorries;
   (e) vibration, from roads or machinery;
   (f) impact loads from boats, vehicles or waves;
   (g) tensional pull of mooring ropes and boats in river and sea wall situations;
   (h) the weight of stacked goods resting on the retained soil.

Types of retaining wall

There are three types of retaining wall:

1. *Gravity walls*
2. *Flexible walls*
3. *Revetments*

1. *Gravity walls* Walls which depend on their dead weight for stability (Figs 67-69). Mostly used for walls up to 2 m high, but may be economically higher if suitable material is available cheaply locally, e.g. stone. Where coursed brick or stone, or uncoursed rubble is used, it is important to avoid tension in the joints. This can occur if a wall is designed as shown in Fig. 70b. It is possible to augment the weight of the wall materials with soil (Fig. 70e) and crib walls (Fig. 71) (open box forms in timber or reinforced concrete

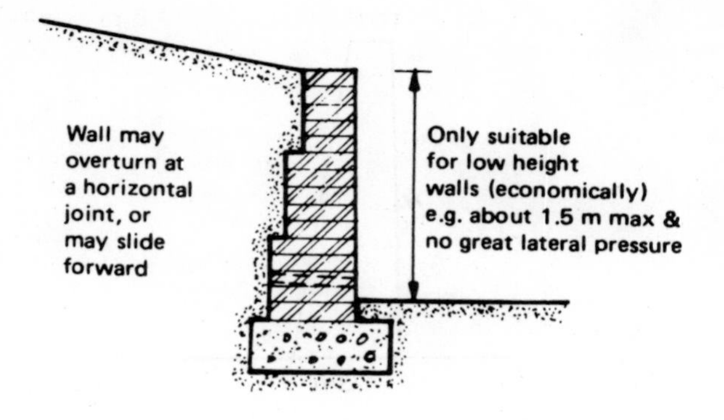

Fig. 67. Mass masonry retaining wall.

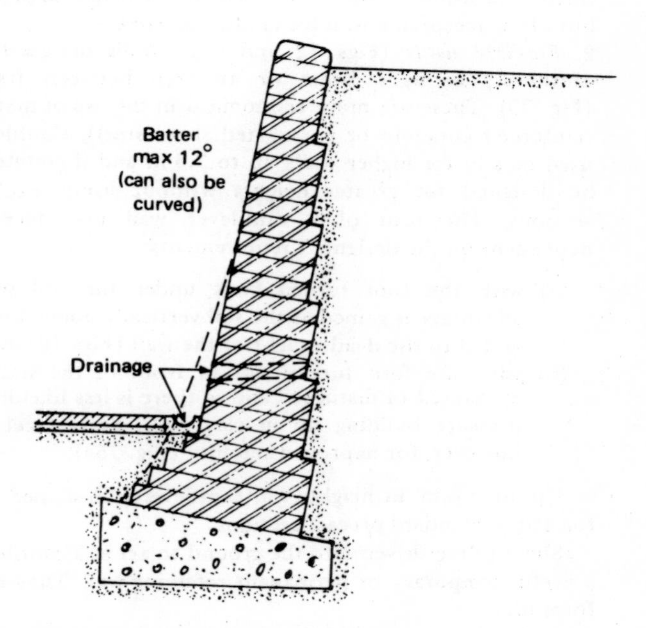

Battered walls are more economical in materials but require more skill in labour

Fig. 68. Battered (inclined) brick retaining wall.

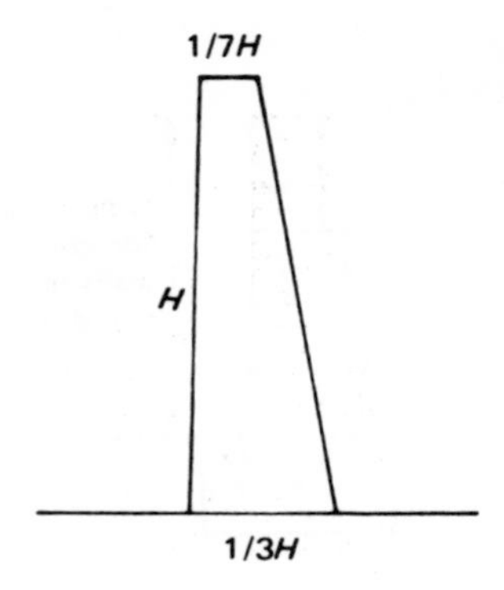

Fig. 69. A guide to initial selection of mass gravity wall proportions.

filled with stones or soil in situ) use this principle to provide a cheap, but often acceptable wall for landscape work.

2. *Flexible walls* (Figs 72 and 66) Walls designed to act as a cantilever, or as a beam or an arch between fixed supports (Fig. 75). These are more economical in the use of material (usually reinforced concrete or corrugated sheet steel). Cantilever walls are used mostly for higher walls up to 7.5 m and if counterforted, may be designed for greater heights without using excessively thick sections. The foot of a cantilever wall may face either way, dependent on the designer's requirements:

(a) with the foot turned back under the soil supported, an advantage is gained as the soil vertically above the foot can be added to the dead weight of the wall (Figs 70e and 75b);
(b) with the foot turned out in front of the wall, less soil is excavated or disturbed and so there is less likelihood of water pressure building up in clay soils. Some heel is advisable, however, for improved stability (Fig. 75a).

Up to 3.6 m in height, r.c. cantilevered L-shaped walls can be formed of standard precast sections.

Sheet piling driven into the ground to act as a cantilever can form a useful temporary or permanent retaining wall. They may be in the form of:

(a) cantilever walls (Fig. 78);
(b) anchored walls with tie rods of steel from the top of the piles to anchors holding against the passive pressure of soil well outside any possible slip planes (Figs 65d and 52).

Anchors can be of:

1. Cast iron or thick steel plates.
2. Large blocks of concrete.
3. Pressure-grouted cement plugs (Fig. 79).

Anchored pile walls can be used up to a height of 11.5 m. Piles may be driven vertical, or angled for better resistance to sliding and impact. A record should be kept of the depth reached by each pile. Drainage holes may have to be cut through the piles, with hinged flaps in sea or river locations.

3. *Revetments* (Fig 80) The surfacing of soil banks or cuttings with masonry or concrete to retain, stabilize and prevent erosion or weathering of the soil. Revetments are easy to build, and economical in materials. Mostly used in landscaping operations and for sea and reservoir walls, as the force of waves is dissipated up the sloping sides (Fig. 81).

Design of retaining walls

The calculations for the design of retaining walls to suit the wide range of possible conditions are beyond the scope of this introductory book on foundations. Civil Engineering Code of Practice No. 2–*Earth Retaining Structures* is a useful guide. The following points are useful however in deciding the form the wall should take:

1. Consider the availability of materials and labour and their relative costs. The cost of shuttering for reinforced concrete can be high, and a simplified shape for casting is desirable on economic grounds (Fig. 76).
2. Consider the most suitable materials and construction for waterproofing where this will be important (as in basements).
3. Consider the restraints that are available or can be made available to assist the wall, e.g. cross-walls, floors, cross-beams in basements (Figs 66 and 86); anchors or ties for piling walls (Fig. 52); weight of soil (Fig. 75b) and post-tensioning (Fig. 77).
4. Consider the use of a curved or folded plan shape to obtain maximum strength with minimum materials (Fig. 73).
5. A downstand at the underside of the base, or taking the foundation further down into the soil will give added resistance to sliding forward (Fig. 72).
6. The more eccentric the loading, the more pressure there will be on the toe of the foundation, tending to cause unequal settlement of the wall and causing it to tilt forward (Fig. 65b).
7. Consider using piles under the wall if the soil is too soft to provide the necessary resistance.
8. Check that there is no possibility of removal of soil from the front of the wall by landscaping work, dredging, scour or other causes.

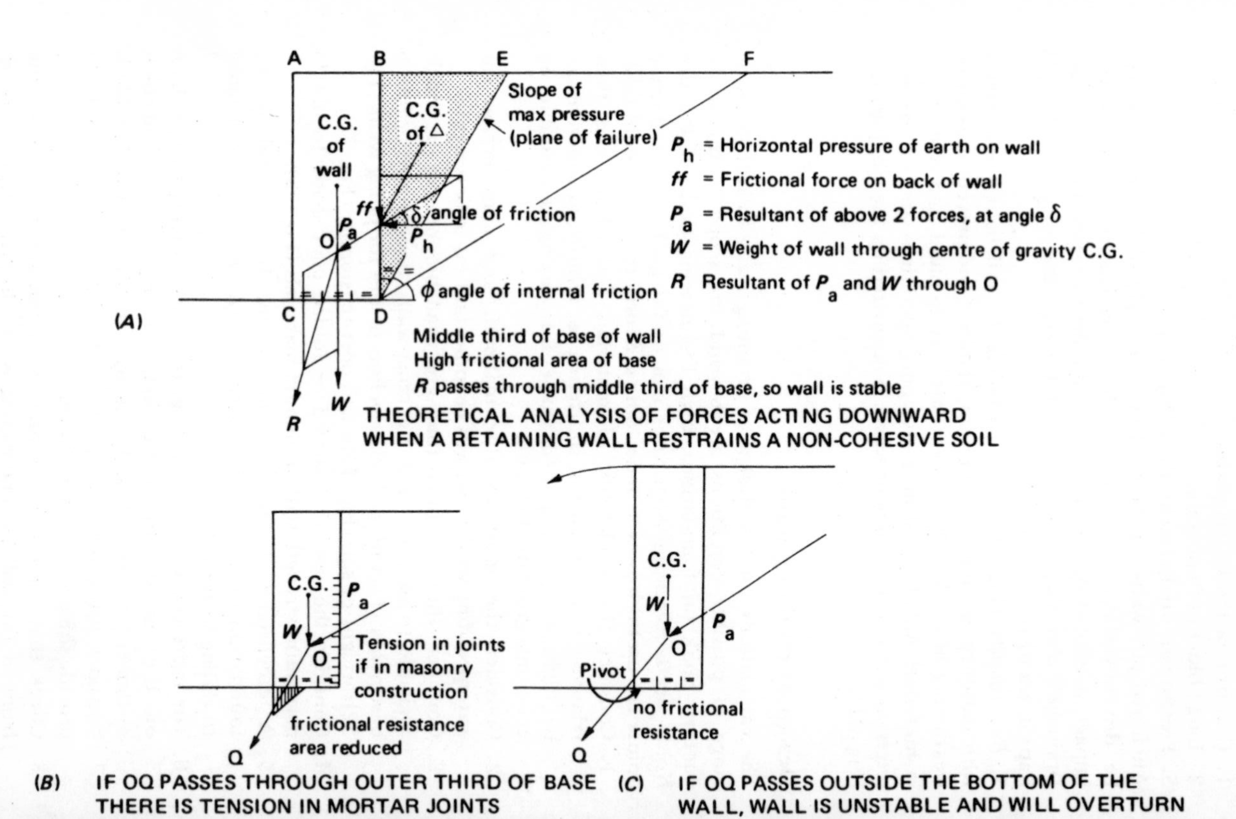

(*A*) THEORETICAL ANALYSIS OF FORCES ACTING DOWNWARD WHEN A RETAINING WALL RESTRAINS A NON-COHESIVE SOIL

(*B*) IF OQ PASSES THROUGH OUTER THIRD OF BASE THERE IS TENSION IN MORTAR JOINTS

(*C*) IF OQ PASSES OUTSIDE THE BOTTOM OF THE WALL, WALL IS UNSTABLE AND WILL OVERTURN

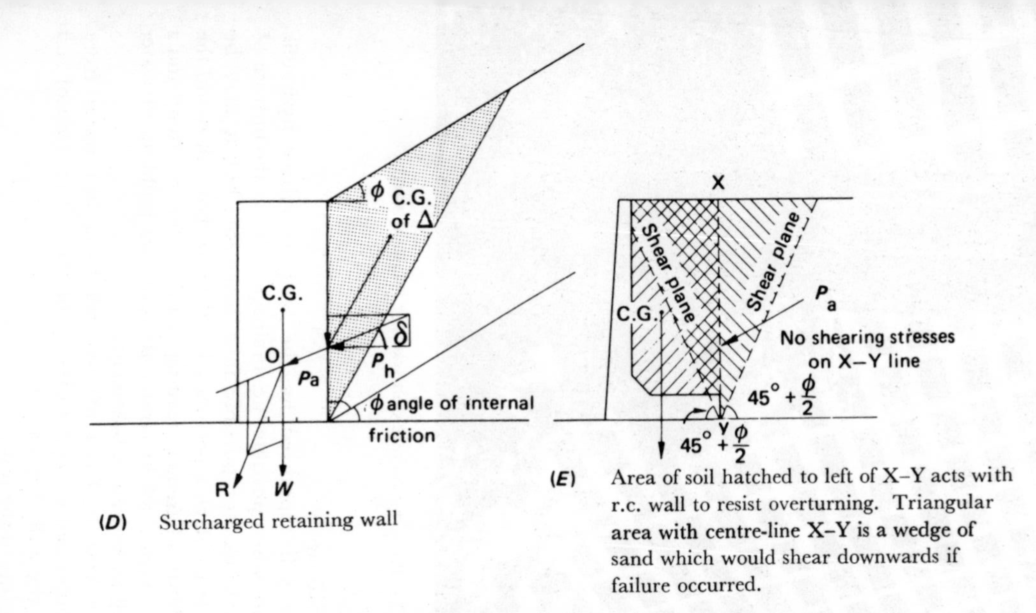

(*D*) Surcharged retaining wall

(*E*) Area of soil hatched to left of X–Y acts with r.c. wall to resist overturning. Triangular area with centre-line X–Y is a wedge of sand which would shear downwards if failure occurred.

Fig. 70. Diagrams of retaining walls for assessment of stability.

Fig. 71. Precast concrete crib wall for retaining earth.

9. Consider drainage:
   (a) of the soil supported by the wall, to relieve hydraulic pressure when possible, i.e. drainage holes at the bottom of the wall are not possible in a basement, but it may be possible to place a perforated collecting drain all round the outside of a basement leading to a drain, or to a sump and a pump. Figure 84 illustrates the pressure build-up of water around an undrained basement:
   (b) of the area in front of the wall to collect water from drainholes and prevent softening of the soil in front of and under the wall (Fig. 68).
10. Consider joints in the wall:
    (a) for construction purposes;

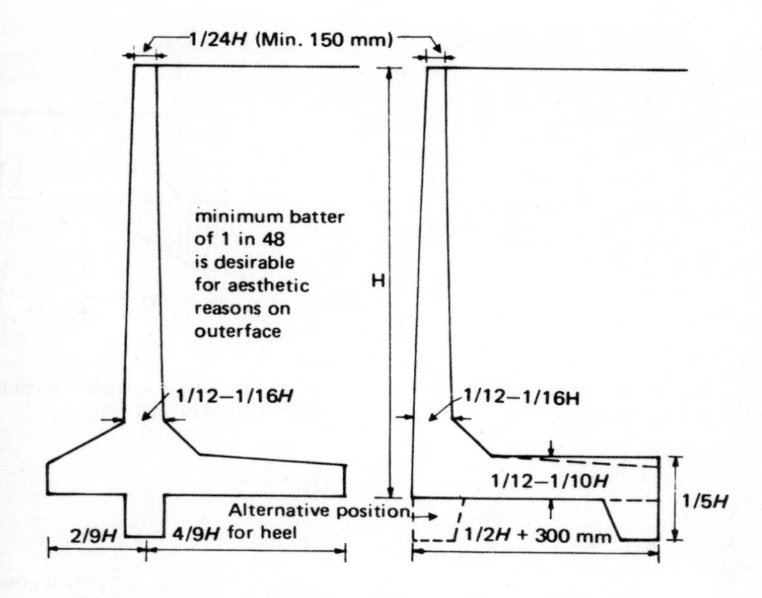

Fig. 72. Trial proportions for reinforced concrete retaining walls.

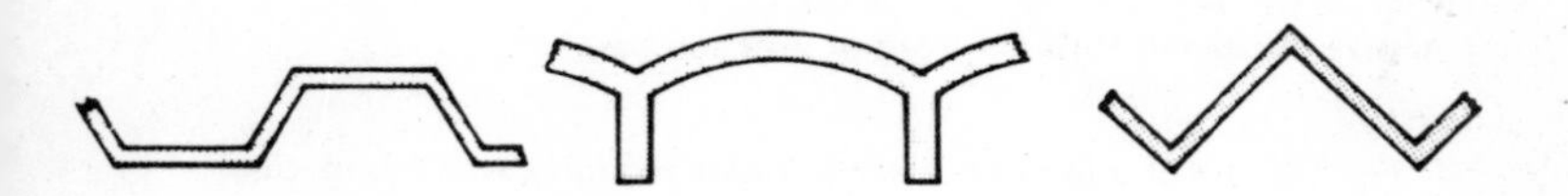

Fig. 73. Alternative plan forms suitable for retaining walls.

(b) for the setting shrinkage of concrete (in long walls a joint is needed every 7.5 m maximum);

(c) for thermal movement (12 mm in every 30 m).

These will present problems when waterproofing is necessary but neoprene jointing strips are available for casting into the concrete at joints (Fig. 86).

11. Consider the penetration of the wall by services, as is commonly required in basements. This may also present problems in waterproofing and in flexibility of connections to take up any movement.
12. Where a choice is possible (as in most cases), consider the appearance of the wall. Select form, material, texture, colour and scale of detail to suit the location and size of the project.

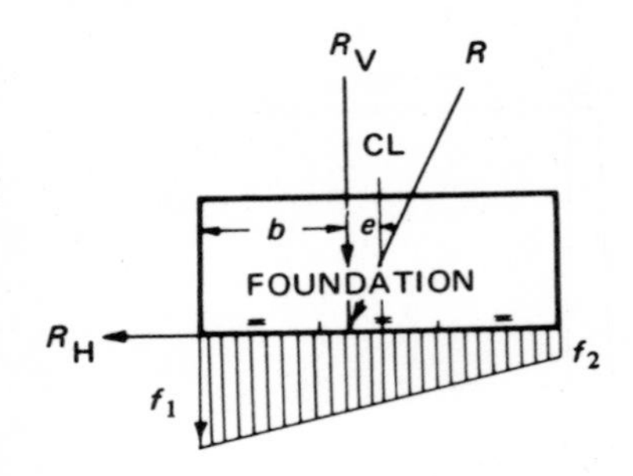

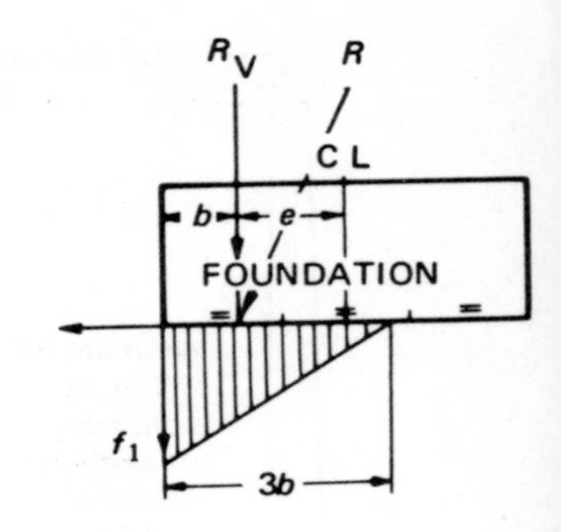

Case 1.

When resultant R is applied within middle third of wall base,

$$f_1 = \frac{Rv}{B}\left(1 + \frac{6e}{B}\right)$$

$$f_2 = \frac{Rv}{B}\left(1 - \frac{6e}{B}\right)$$

Case 2

When resultant R is applied outside middle third of wall base,

$$f_1 = \frac{2}{3}\,\frac{Rv}{b}$$

$$f_2 = 0$$

$\therefore$ risk of failure of soil at f_1 is greate

Note: R can be resolved into R_V vertical force and R_H Horizontal force, the latter tending to cause horizontal sliding forward, resisted by:

(i) friction between concrete base & soil under
(ii) passive resistance of soil in front of wall and base.

Fig. 74. Distribution of pressures from a retaining wall foundation.

Construction of retaining walls

The major problems in the construction of retaining walls are:

1. *Retention of soil* while the wall is being built, when excavation is necessary.
2. *Control of water.*
3. *Control of the quality* of materials and workmanship.
4. *Careful backfilling.*

Retention of soil while excavating will be described in a later chapter under "Excavation". Some of the methods of controlling water on site will also be mentioned. Retaining walls, however, need

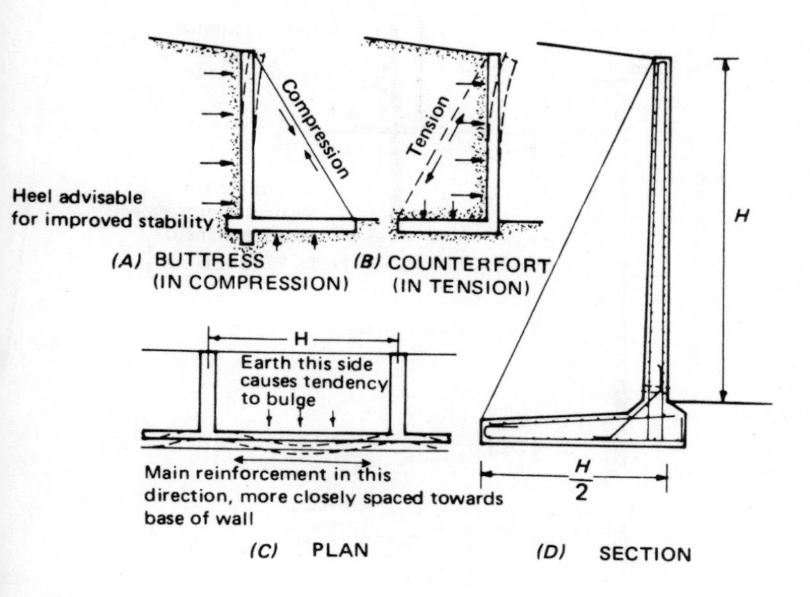

Fig. 75. Reinforced concrete retaining wall with counterforts, for heavily loaded high wall.

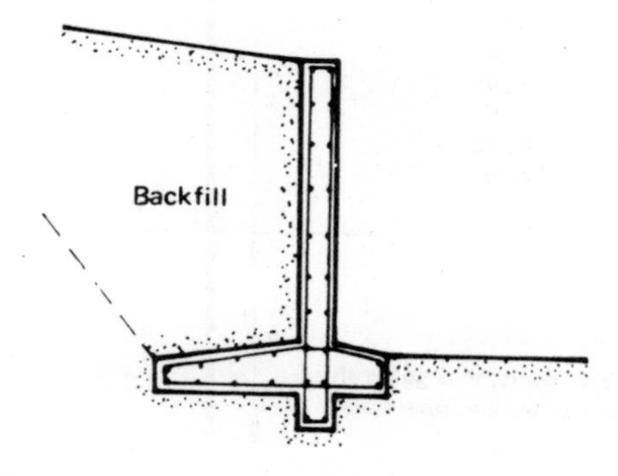

Fig. 76. Reinforced concrete retaining wall (simple section for easy shuttering).

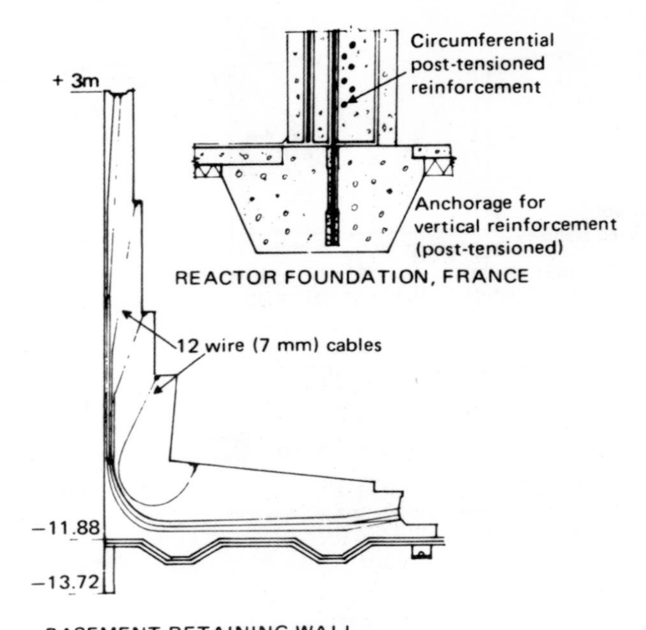

Fig. 77. Post-tensioned reinforced concrete foundations.

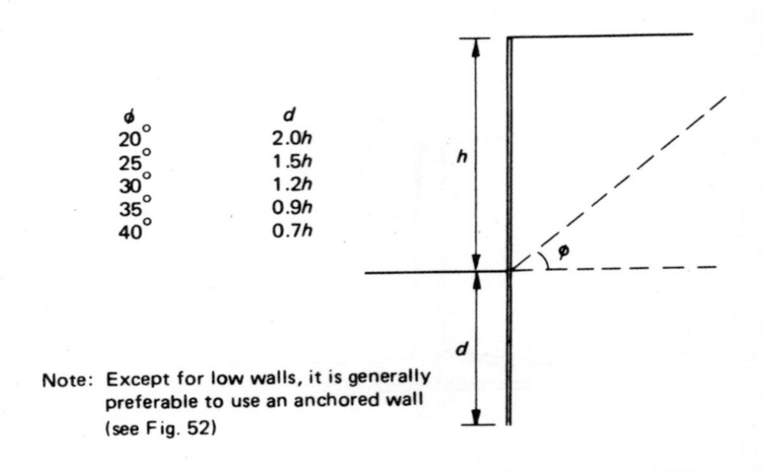

ϕ	d
20°	2.0h
25°	1.5h
30°	1.2h
35°	0.9h
40°	0.7h

Fig. 78. Approximate depth in non-cohesive soils necessary for cantilevered piling acting as retaining wall.

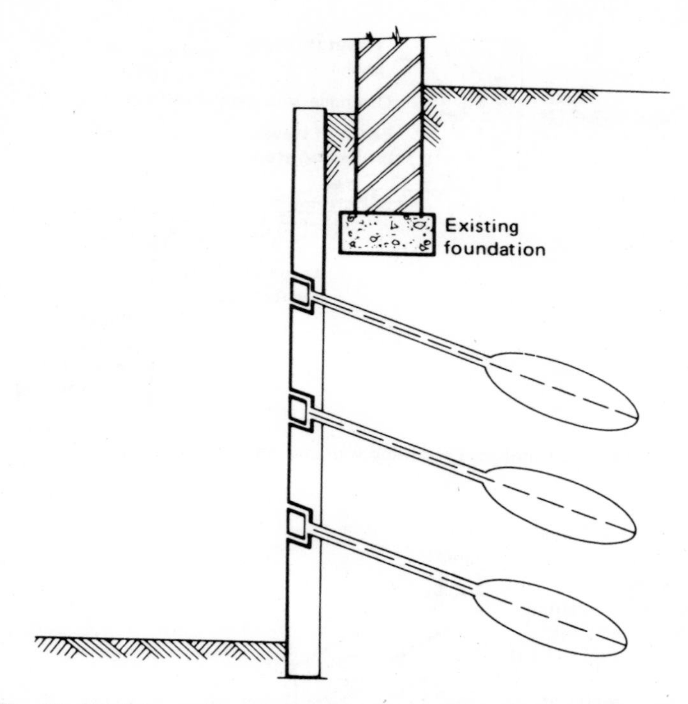

Fig. 79. Retaining wall anchored with reinforced pressure grouting giving uncluttered space for excavation.

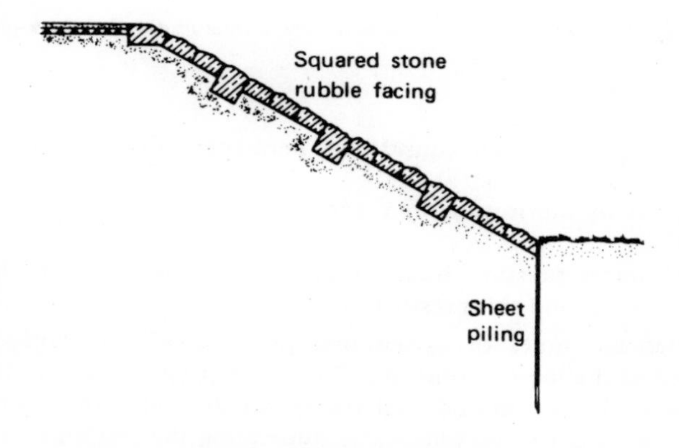

Fig. 80. Typical revetment.

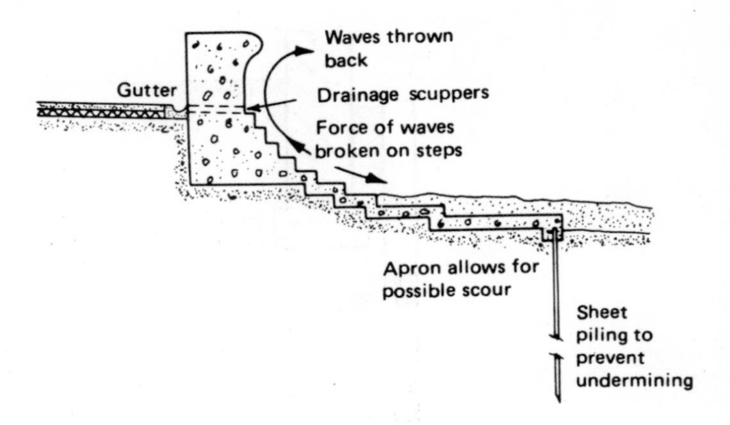

Fig. 81. Combined retaining wall and revetment for sea wall.

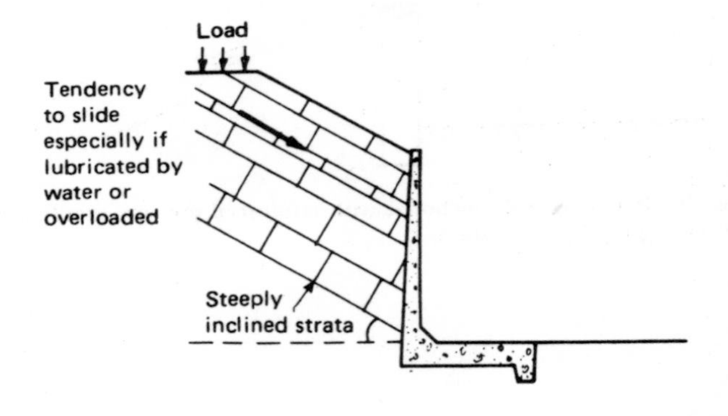

Fig. 82. Steeply-sloping strata may need a retaining wall or rock-pinning.

particular care when considering permanent drainage, for failure often occurs through:

(a) water lubricating slip planes

or

(b) water pressure building up behind a wall not designed to withstand this pressure.

Drainage holes or agricultural pipes should be fairly closely spaced at the lowest points possible, permitting drainage to the front of the wall. They should discharge into a channel or on to a concrete apron designed to prevent water lubricating the soil under the wall, and thereby causing a slip.

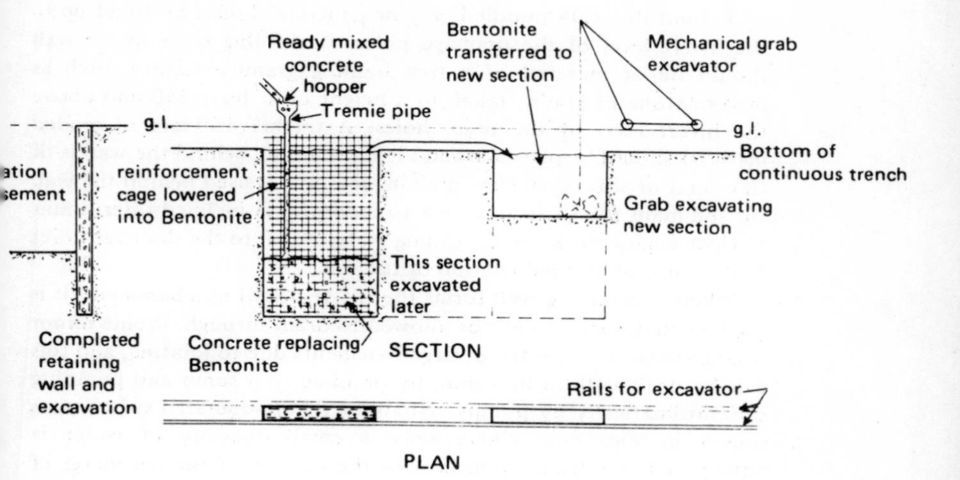

Fig. 83. Construction of retaining wall by temporary use of thixotropic suspension to maintain stability of the trench sides.

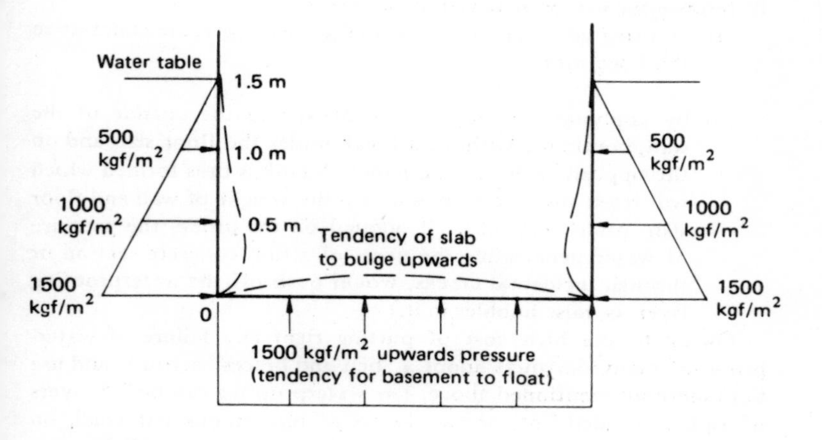

Weight of concrete = 2240 kg/m^3

∴ 670 mm thickness of concrete would be necessary to counteract upward pressure of water at 1.5 m depth below water table, or r.c. slab spanning between basement walls could be used to reduce thickness. (Design as an inverted slab with evenly distributed load and place tension reinforcement at top of slab.)

Fig. 84. Water pressure on basement below water table.

Behind the wall, puddled clay or concrete should be filled up to the invert level of the drainage holes. Backfilling close to the wall should be of material of a free-draining granular nature, such as broken stone or gravel, taken to a height of at least 450 mm above the invert level of the drain holes. A further 150 mm of graded material should be placed around it if the filling behind the wall is of fine sand or silt. When slow draining material is used behind the wall for the main filling, it is as well to incorporate inclined filter drains at right angles to the wall, leading water down to the drainage holes at the angle of internal friction of the soil.

Where a retaining wall forms the external wall of a basement, it is obvious that water cannot be allowed to drain through. (Some minor leakage must be expected in large basements due to jointing, and this can be dealt with as in a ship, by draining to a sump and pumping out automatically by pumps actuated by float-operated switches, as shown in Fig. 86). Where only a small quantity of water is anticipated, a porous drain around the outside of the perimeter of the wall at its base can collect up the water and drain it away to a convenient surface water drain or soakaway well. When, as in many cases, there is a high water table around the basement, then it must be designed to withstand the considerable water pressure, and must also be waterproof under that pressure.

Waterproofing can be achieved in two ways:

(a) by using good quality dense well-graded aggregate concrete in thick sections

or

(b) by applying a waterproof membrane to the *outside* of the wall, continued without a break under the floor slab and up the opposite side of a basement. A tank is thus formed which will resist the water pressure by the weight of wall and floor slab inside (Fig. 85). (If applied on the *inside,* the pressure of water penetrating the pores of a thin concrete section or through incidental cracks, would push off the waterproofing layer, or raise bubbles in it.)

Owing to the high cost of putting right any failure of waterproofing, many designers adopt a "belt and braces" attitude and use both methods mentioned above. The waterproofing can be 2-3 layers of asphalt applied hot, or two layers of bituminous felt stuck on with bitumen and laid to break joint. Failures in such membranes are often very difficult to trace to the exact source.

Materials for the wall itself are usually chosen from:

1. Concrete (reinforced as necessary).
2. Brick.
3. Stone.
4. Precast concrete blocks.

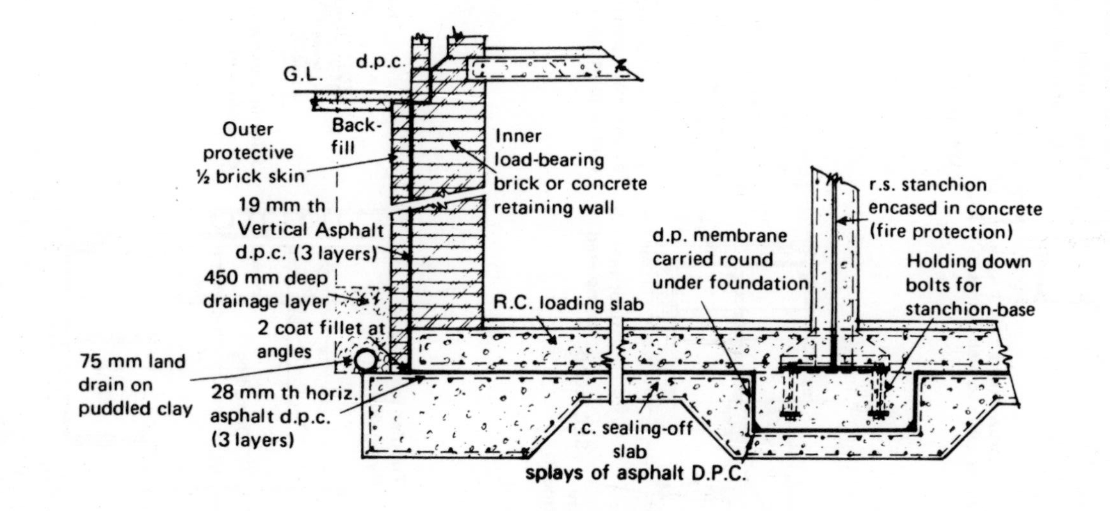

Fig. 85. Section through small basement with R.S.S. base and foundation.

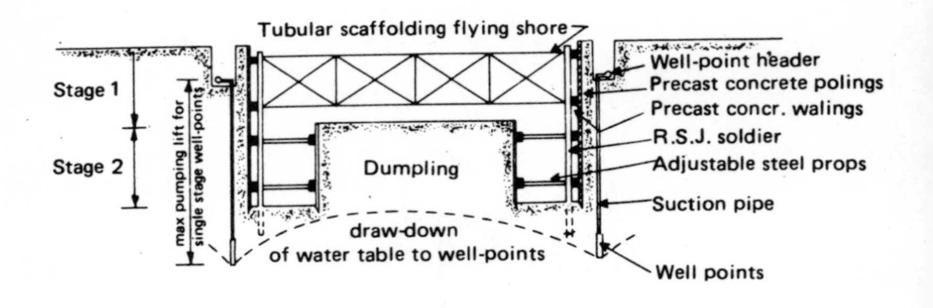
Tubular scaffolding flying shore
Stage 1
Stage 2
max pumping lift for single stage well-points
Dumpling
draw-down
of water table to well-points
Well-point header
Precast concrete polings
Precast concr. walings
R.S.J. soldier
Adjustable steel props
Suction pipe
Well points

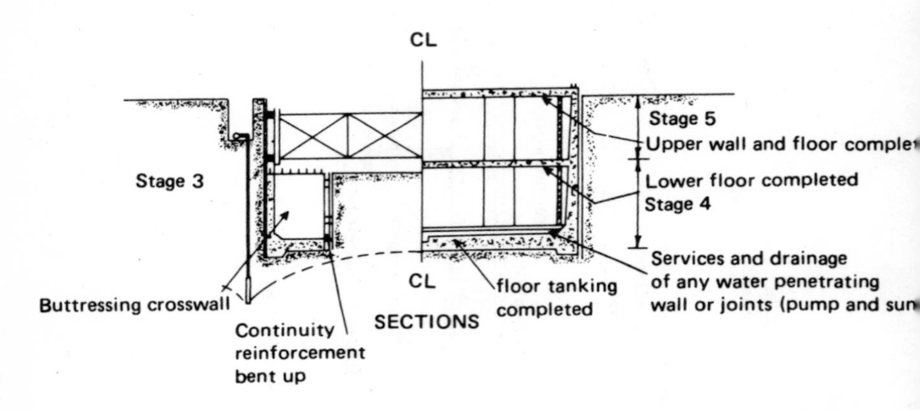
CL
Stage 3
Stage 5
Upper wall and floor comple
Lower floor completed
Stage 4
Services and drainage
of any water penetrating
wall or joints (pump and sum
CL
floor tanking
completed
Buttressing crosswall
Continuity
reinforcement
bent up
SECTIONS

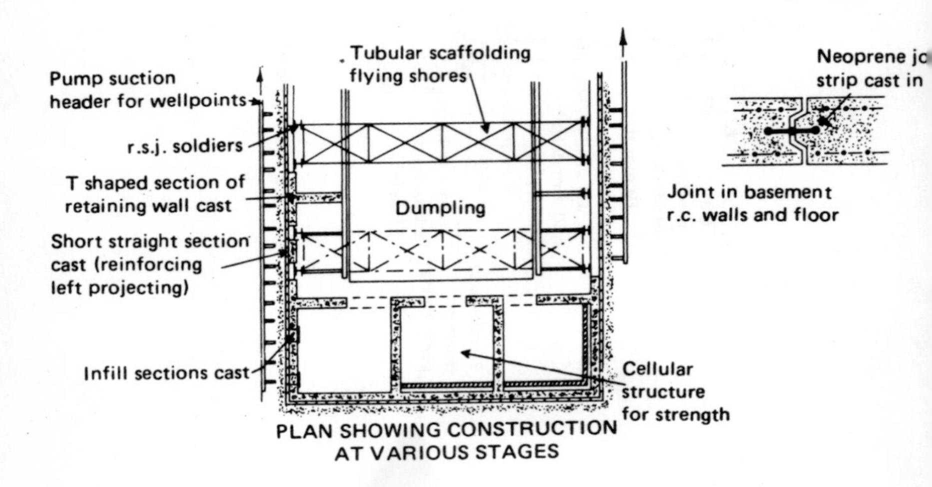
Tubular scaffolding
flying shores
Pump suction
header for wellpoints
r.s.j. soldiers
T shaped section of
retaining wall cast
Dumpling
Short straight section
cast (reinforcing
left projecting)
Infill sections cast
Cellular
structure
for strength
PLAN SHOWING CONSTRUCTION
AT VARIOUS STAGES
Neoprene jo
strip cast in
Joint in basement
r.c. walls and floor

Fig. 86. Construction of deep basement.

Concrete for mass-retaining walls need not be of stronger mix than 1 : 4 : 8 except where such walls must be waterproof, resist tension, or have a good finish (and this last requirement can be achieved with only a thin surface layer). For waterproof concrete, a better mix is desirable, the thinner the section; and no mix weaker than 1 : 2 : 4 is likely to be effective. A very well-graded aggregate is essential, well mixed and well compacted. There is no room for bad workmanship. Should patchiness occur, allowing water to penetrate, it can best be remedied first by drilling holes in the centre of each patch and then pressure-grouting.

In this country, bricks used for retaining walls must be frost-proof and engineering quality bricks are preferable. Since moisture dries out on the exposed face, an efflorescence-free brick should be used for the sake of appearance. Mortar mix should be 1 : 3 cement and sand, and if the wall is supporting sulphate-bearing clay, high alumina or other sulphate-resisting cement should be used in place of the usual Portland cement.

Stone must also be frost-resistant; granite and the harder sandstones being most suitable for larger walls.

Concrete blocks must be of dense concrete, and their finish should be carefully considered.

Further references on retaining walls, basements and waterproofing

Civil Engineering Code of Practice No. 2
Earth Retaining Structures — Institute of Structural Engineers

Application of Mastic Asphalt—Tanking and Damp-proof Coursing
The Natural Asphalt Mine-owners and Manufacturers Council

Stressed Ground Anchors—
Paper by Dr. K. G. Stagg — University of Bristol Symposium on New Developments in Foundation Design

Chapter 9—Some special problems in foundation design

Once he or she has learned how to tackle normal foundation design, the designer should be equipped to analyse the special problems that arise from time to time, and to arrive at a satisfactory solution. Some example of these problems are touched on briefly below.

Pre-fabricated factory-made buildings and structural frames

Much attention has been paid in recent years to the transfer of as many building operations as are possible to the factory, where industrial methods can be used to bring down costs and provide better conditions for the workers. Unfortunately, little has been done, and in many cases little can be done, to improve the construction methods normally employed for foundations. Some lightweight buildings can use pre-fabricated short piles or r.c. pads and some building systems have experimented with these, notably the "CLASP" system (Figs 88-91). Other systems such as "IBIS"

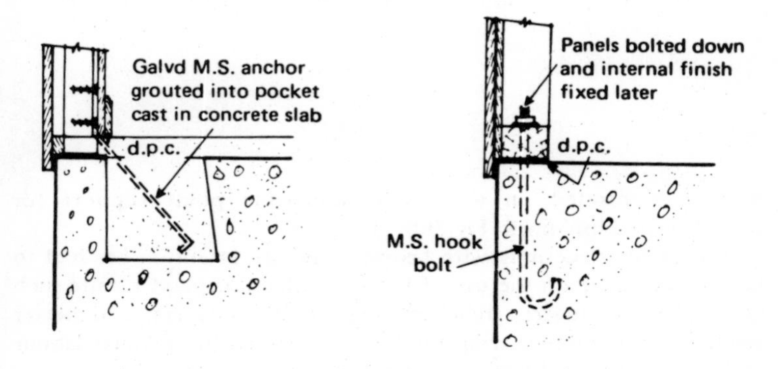

Fig. 87. Alternative methods of anchoring prefabricated timber-framed wall units to concrete foundation slab.

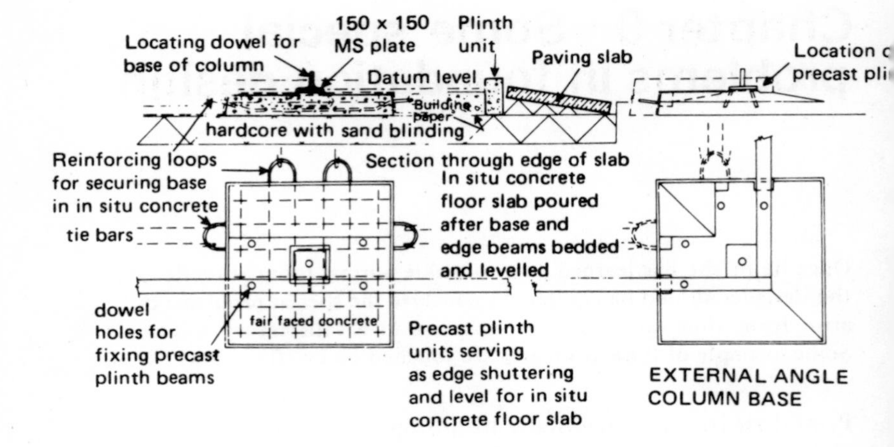

Fig. 88. Precast column base as used in CLASP construction.

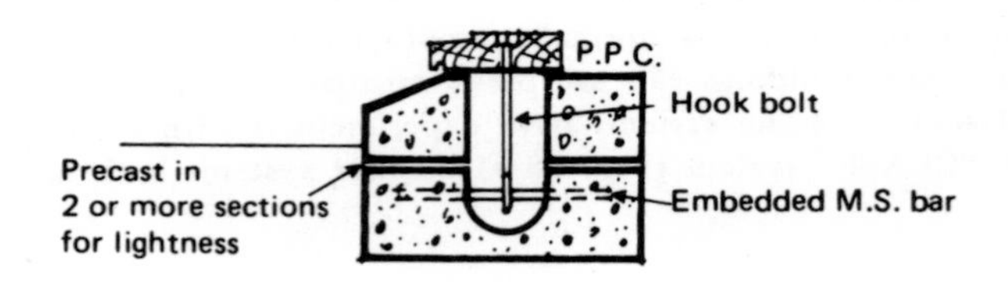

Fig. 89. Precast concrete foundation blocks for light construction.

have concentrated on a suitable adjustable point support for pre-fabricated buildings (Fig. 92).

For larger structures with heavier loadings, attention has had to be focused more on the use of better mechanical plant on site such as mechanical augers, trenchers, diggers and scrapers; and better earth- and concrete-moving methods to reduce the manual labour and unpleasant conditions in all weathers on site.

Joints transferring loads from pre-fabricated structural members to the foundations are special problems for which some solutions are shown in Figs 43-46.

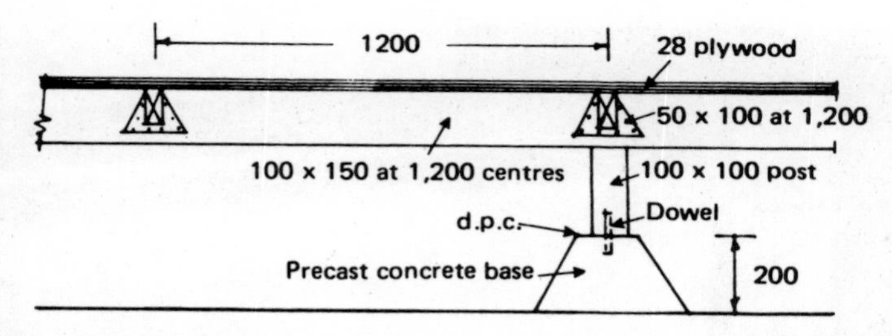

Fig. 90. American design for a precast concrete base for a suspended timber ground floor.

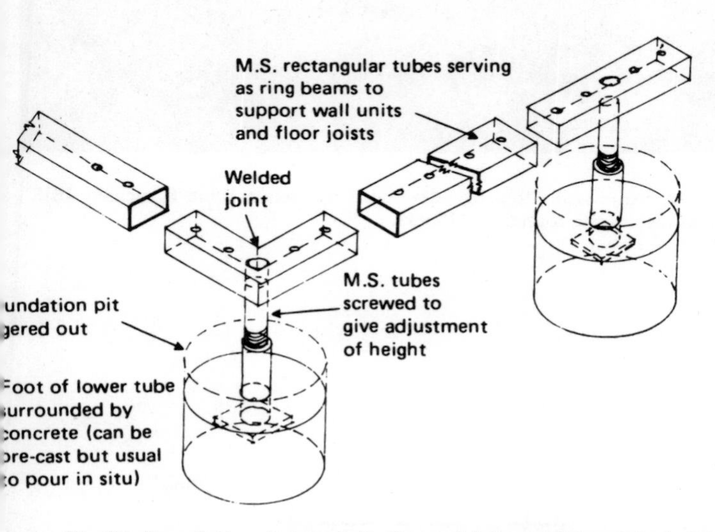

Fig. 91. Foundations in bored pits for prefabricated lightweight building.

Foundations on site boundaries

A common problem in urban areas is the building with structural columns occurring along the line of a boundary. This results in a situation where a normal pad foundation would be overlapping neighbouring property. Two possible solutions exist, the choice being dependent on which best suits the particular conditions. One is the balanced foundation for two adjacent columns shown in Fig.94, and the other is the cantilevered beam foundation shown diagrammatically in Fig. 95.

Fig. 92. Adjustable footing: developed to connect pre-fabricated IBIS system houses to site foundation blocks.

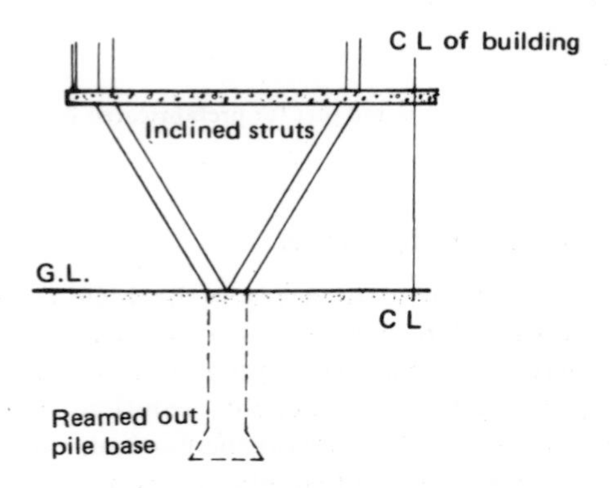

Fig. 93. Combined foundation for two columns, supported on one large-diameter bored pile.

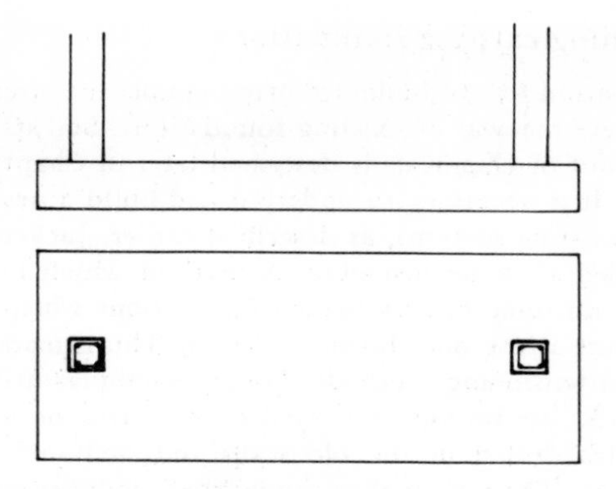

Used where columns occur at site boundaries
& must sit eccentrically on foundations. Combining
the 2 foundations gives balance

Fig. 94. Combined foundation.

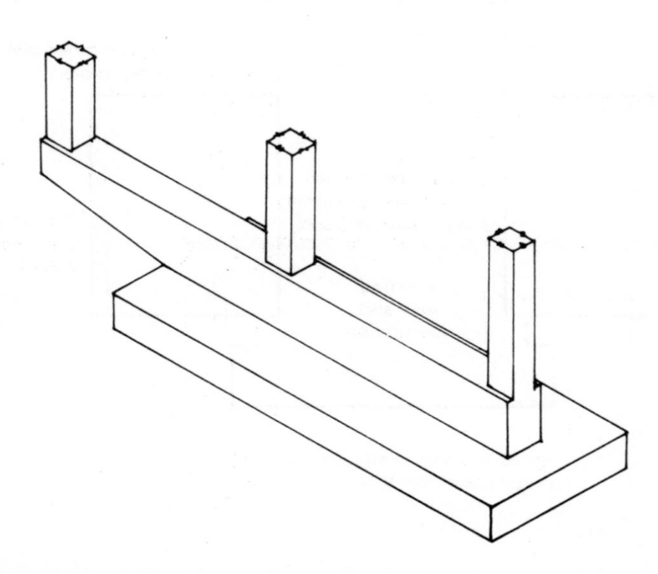

Fig. 95. Cantilever foundation, to avoid pressure of new structure on old sub-structure adjacent.

Strengthening existing foundations

The preservation of old buildings often entails the strengthening or even complete renewal of existing foundations. Soil stabilization by injecting grout or chemicals is described later in Chapter 10, but in many cases it is necessary to underpin and build a new foundation under the existing sections, as described earlier. Jacked-in sectional piles are used in some instances. A method which has been used successfully on some heavily loaded foundations which have shown signs of failure is the one shown in Fig. 96. This is made possible by modern post-tensioning methods. These techniques utilize specially designed jacks to tension and wedge the wires or cables passed through holes drilled in the old walls and anchored in pile head anchor blocks. The wall is thus sandwiched under pressure and the load transferred to the new piles formed on each side in pairs.

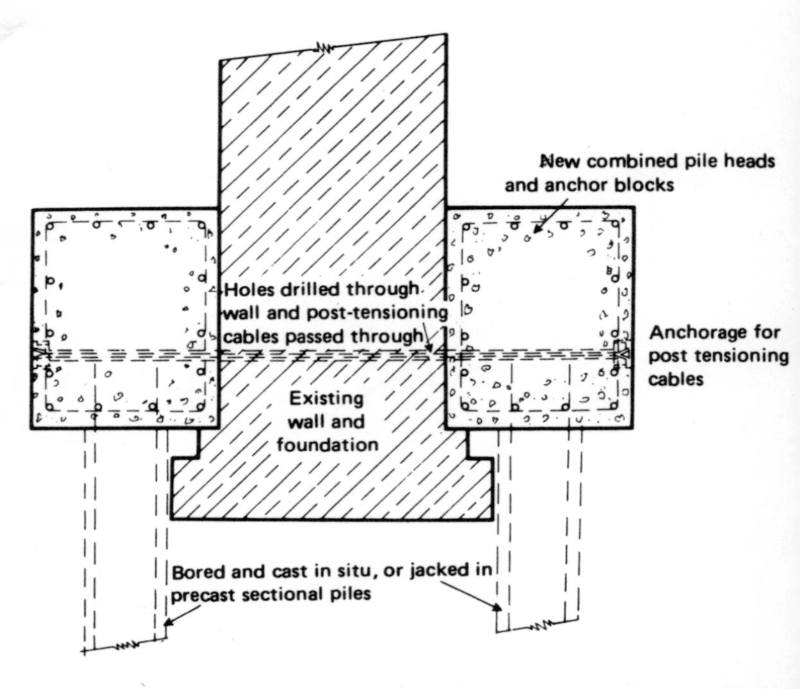

Fig. 96. Strengthening of old foundations: paired piles with pile heads linked by post-tensioned cables through old wall.

Sound and vibration insulating foundations

Mention has already been made of the dangers of vibration transmitted through foundations from machinery. Such vibration can cause failure through settlement resulting from consolidation of the soil beneath. There is also the nuisance of noise and vibration being transmitted throughout a building from internal or external sources. Figure 97 shows a typical solution to this type of problem.

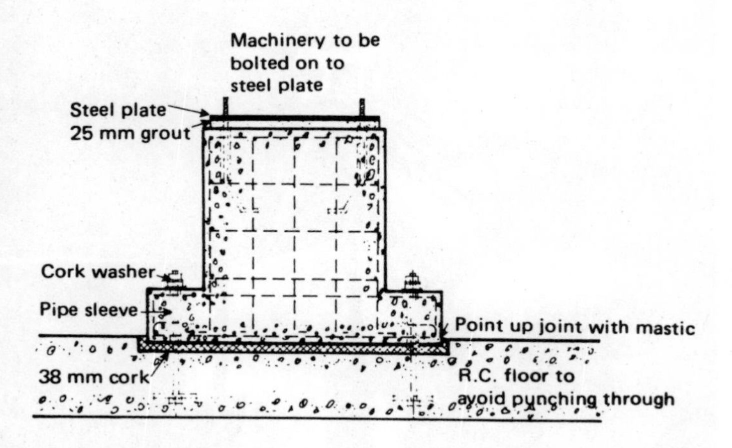

Fig. 97. Foundation for machinery liable to cause vibration and noise requiring insulated bed and vibration-proof nuts. Floor slab and bolts must resist shear.

Foundations subject to tension

There is greater awareness today of the potential economy of material in tension structures, and of the exciting new forms possible where the majority of the structural members are in tension. Developments in materials and techniques are resulting in tent-like structures, whose roofs are hung in catenary curves or stretched over tensioned cables. Most cable structures need end anchorages which tend always to be pulled out of the ground, and are therefore the reverse of the traditional foundation. Careful analysis of the magnitude and direction of these tensional forces is necessary and new thought has to be given to methods of securing the foundation in the ground.

Fig. 98. Reinforced concrete anchorage for tensioned tent-roof structure, W. German Pavilion, Expo 1967, Montreal.

A typical example of this type of structure was the West German pavilion at Expo 1967, Montreal. This had a tent roof supported on masts and wire cables tensioned back to reinforced concrete end anchorages of most effective design (Fig. 98). In rock, tensioned cables can be secured by drilling holes and grouting in cement and sand somewhat as in Fig. 79. Alternatively, they may be post-tensioned to concrete abutments secured to the rock.

With other foundation conditions, there may be greater difficulty. Frictional piles with reamed out or bulb ends can be used in stiff

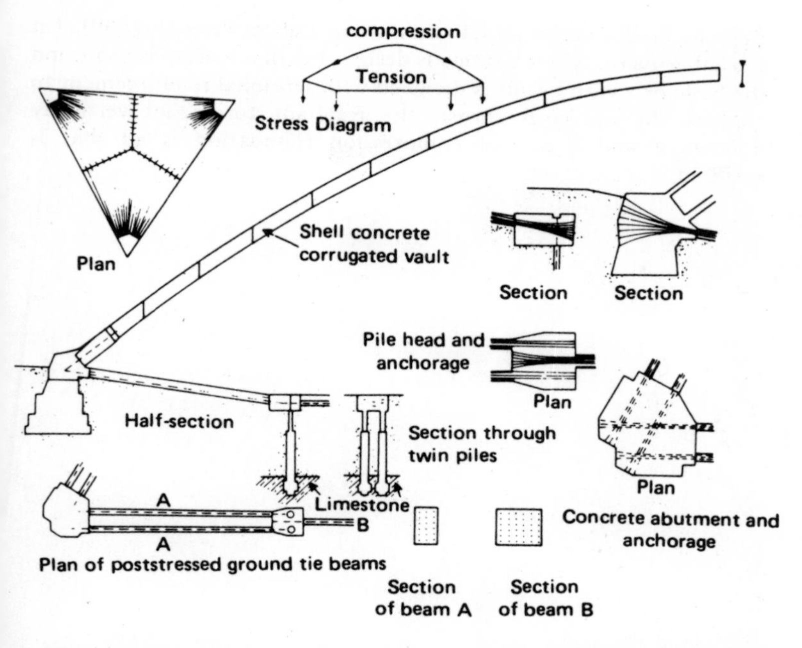

Fig. 99. Complex foundation used in C.N.I.T. Exhibition Hall, Paris: tension members are provided to prevent outward thrust of huge shell vault (spanning 205.5 m, rise 46.3 m).

clay or firm sand, if taken deep enough to withstand the pull-out forces. On softer soils, anchorages forming part of the perimeter of a reinforced concrete raft or basement supporting the compression members and forming the floor of the building would appear to be a feasible solution, but expensive unless justified by other requirements of the brief.

There is a warning here that while almost anything is structurally possible today, not everything is economical, and maybe a different type of structure would need to be used in these circumstances.

With the various types of ultra-lightweight, pneumatically inflated structures now available, the structure itself tends to lift and move sideways under wind pressure and suction. These "air structures" are therefore usually weighted down at ground level with integral tubes around the perimeter filled with water or sand, but in addition are pegged down all round to the ground in the same way as a tent.

Buildings in which there are unresolved outward thrusts at ground level (as with many domed and vaulted forms) may have heavy buttress foundations or angled piles to take the load. In some cases however, they may be secured by tensioned ties under the ground floor as in the huge C.N.I.T. Exhibition Hall in Paris (Fig. 99). On the other hand, where a dome is designed with a tension ring around the base, or a barrel vault is designed with tensional reinforcement to contain the outwards thrust, the resultant forces act vertically downward and a normal compression foundation is all that is required.

Chapter 10—Preparing the site for foundations

The area of site to be used for building must first be cleared of all vegetation, with particular attention to tree roots and stumps. Such is the force behind living vegetation, that even grass can force its way through concrete. Although tree roots are dangerous under buildings and some trees such as poplars with wide spreading surface roots should not be allowed to grow within 5 m of foundations, as they dry out the soil and cause differential settlement and cracking; a plea is made for the retention of existing trees and shrubs on the plot whenever possible. Some builders will bulldoze the lot indiscriminately. Good quality turf can be stored for re-use later, and also the loam, or top spit as it is called, which must be removed, usually by mechanical means, to get to firm undisturbed soil. Existing drains must also be removed and diverted.

Having cleared the approximate area of the building, accurate setting out is required. Figure 100 illustrates the method used for a simple structure. Accuracy is of the utmost importance at this stage, as even the smallest errors can be accumulative.

Inaccuracies can occur in measurement by:

1. Inaccuracy of the tape—a linen tape gives an accuracy of between 1/50-1/100, a steel tape of between 1/500-1/1000.
2. Sag of measuring tape.
3. Sloping tape.
4. Insufficiently straightened tape.
5. Inaccurate alignment.
6. Temperature: variation of length due to a change of 28°C gives 1/3000 error with a steel tape.
7. Over-tensioning of tape.
8. Marking position of end of tape out of vertical, especially if a plumb-bob is used to transfer measurements vertically.
9. Thickness of chalk lines.
10. Inaccuracy of optical instruments:

Optical square	± 5 mm in 15 m
Cowley level	± 5 mm in 30 m
Theodolite (graduated to 20 seconds)	± 10 mm in 50 m

See B.R.S. Digest (2nd Series) No. 114 *Accuracy in Setting Out* (H.M.S.O.) for further reference on this subject.

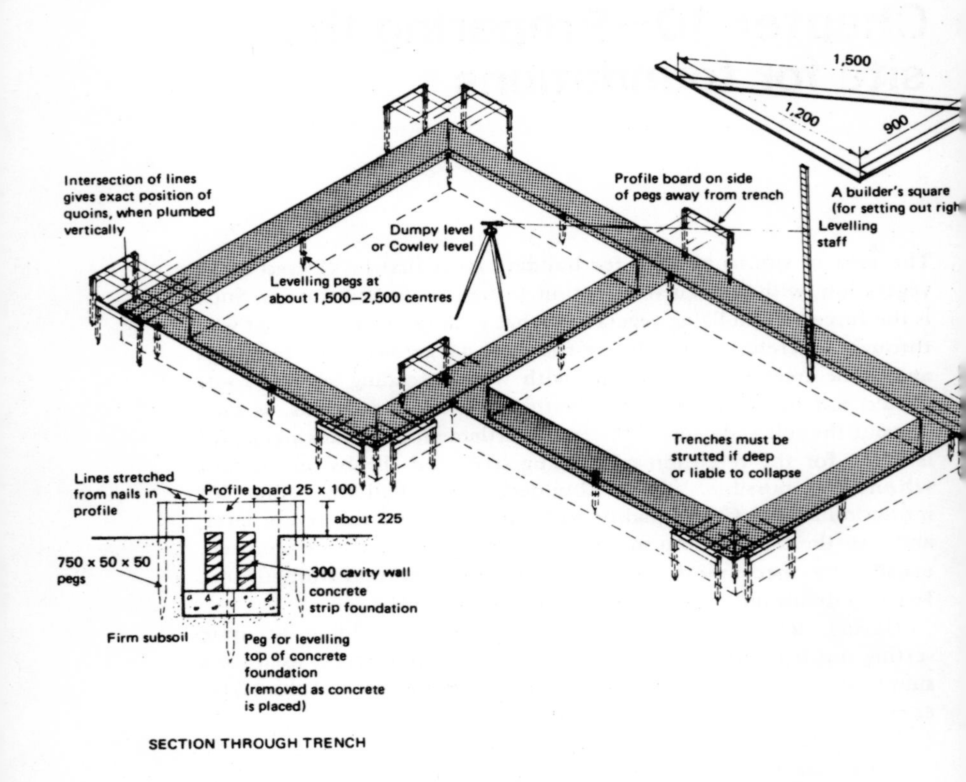

Fig. 100. Setting-out of foundations.

A B.R.S. survey has confirmed the low standards of the building industry in this aspect, with inaccuracies of up to 50 mm in 4.5 m spacing of columns. This is disastrous for fitting pre-cast concrete infill panels. Setting out by the builder should be checked by the clerk of works, or if there is not one on the site, by the architect. A datum level must be established, related to an Ordnance Survey level provided by some nearby Bench Mark, and situated where it will not be disturbed by building operations. Such a datum is preferably a steel or concrete post, set firmly in a concrete base with the top at the chosen datum level. On a small job a wood stake is sometimes used.

From this datum level, all vertical measurements up and down are made. The use of optical measuring equipment such as levels, theodolites, optical squares and optical plumb sights is beyond the scope of this book, but a suitable reference is *Basic Metric Surveying* by W. S. Whyte (Butterworth) which also shows how to set out those radiused and free-form curves which look fine on the drawing but create considerable problems in setting out on the site.

Excavation

Whereas there will always be some labour with hand tools for excavation, the majority of the work is now done by mechanical means, using petrol or diesel oil-driven engines on caterpillar-tracked or large pneumatic-tyred wheel machines. Type of machine include:

Scrapers, bulldozers, face shovels.
Toothed rippers.
Excavators with jib or boom, with toothed bucket.
Excavators with dragline bucket.
Trench excavators with small-toothed buckets on a continuous belt.
Cranes with toothed grabs.
Mechanical augers (for large diameter holes).
Trucks for removing spoil from site.
Dumpers for moving excavated soil and depositing it elsewhere.

It is as well to remember that suitable access to the site is required for these large machines, with sufficient hard-standing to prevent them from getting bogged down in winter and turning the site into a sea of churned-up mud.

Excavating in rock may be possible with mechanical equipment. Certainly pneumatic drills and rotary drills can be used, the latter for providing holes for explosives in hard stone. Explosives can be expensive as well as dangerous and their use may lead to injunctions from neighbouring owners, or complaints about noise in town sites. However, a great deal of rock can be excavated without recourse to explosives by splitting it out with steel wedges at natural bed and fissures. Softer stone can often be ripped out with the powerful toothed rippers now available.

Soft soils are easy enough to remove, but may present the problem of restraining collapse of the sides of excavations. For the safety of the men working on the site, it is essential to provide support of a suitable nature, and indeed it is unlawful not to do so (see the *Building (Safety, Health and Welfare) Regulations 1948 Pt. IV, Sections 75-78)* where the excavation is more than 1.2 m deep unless it can be proved there is no risk. The diagram in Fig. 10 will emphasize the increased risk incurred when excavated material or other building material is piled along the edge of a trench.

Support in excavations can be provided temporarily by:

(a) *Timbering*—Vertical planks or baulks are driven into the ground at the sides of the excavation as it proceeds and held vertical by horizontal timbers called walings, strutted across the trench width by either pit props or similar timber sections, or by adjustable steel props (Fig. 101). Owing to the need for this cross strutting, it is usual when excavating basements to leave the centre "dumpling" unexcavated while the outside walls are being constructed, so as to afford a suitable resistant surface (Fig. 86).

(b) *Steel sheet piling*—either driven in to form a cantilever or more usually with walings and strutting as above. Sometimes the piling is left in position permanently, but it may be economic to withdraw it once a permanent wall has been built.

(c) *Pre-cast concrete piling.* This is usually left in place as permanent shuttering to the vertical wall of a basement (Fig. 86).

(d) *Raking shores* can be used in deep excavations where these are excavated in short sections at a time. They tend to get in the way of work on the floor of a basement (Fig. 102).

(e) *Flying shores* in timber up to about 7.5 m span and in braced steel scaffolding over this up to about 18 m (Figs 86 and 102b).

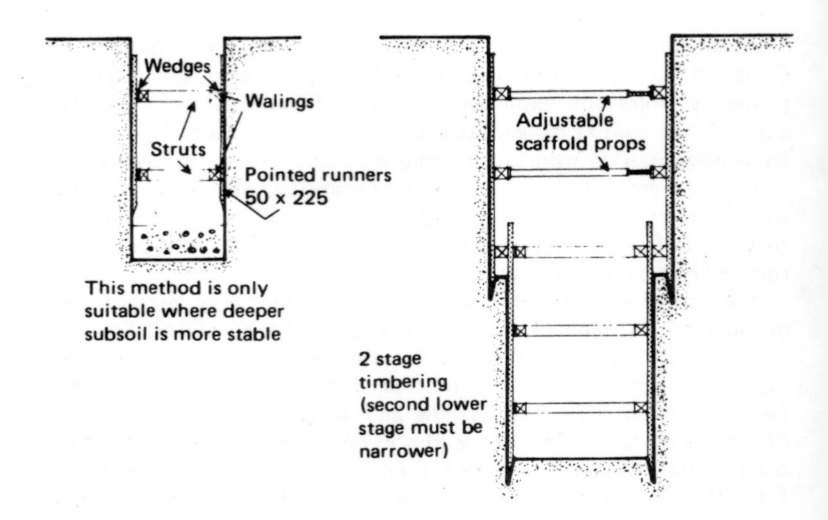

Fig. 101. Timbering of deep trenches.

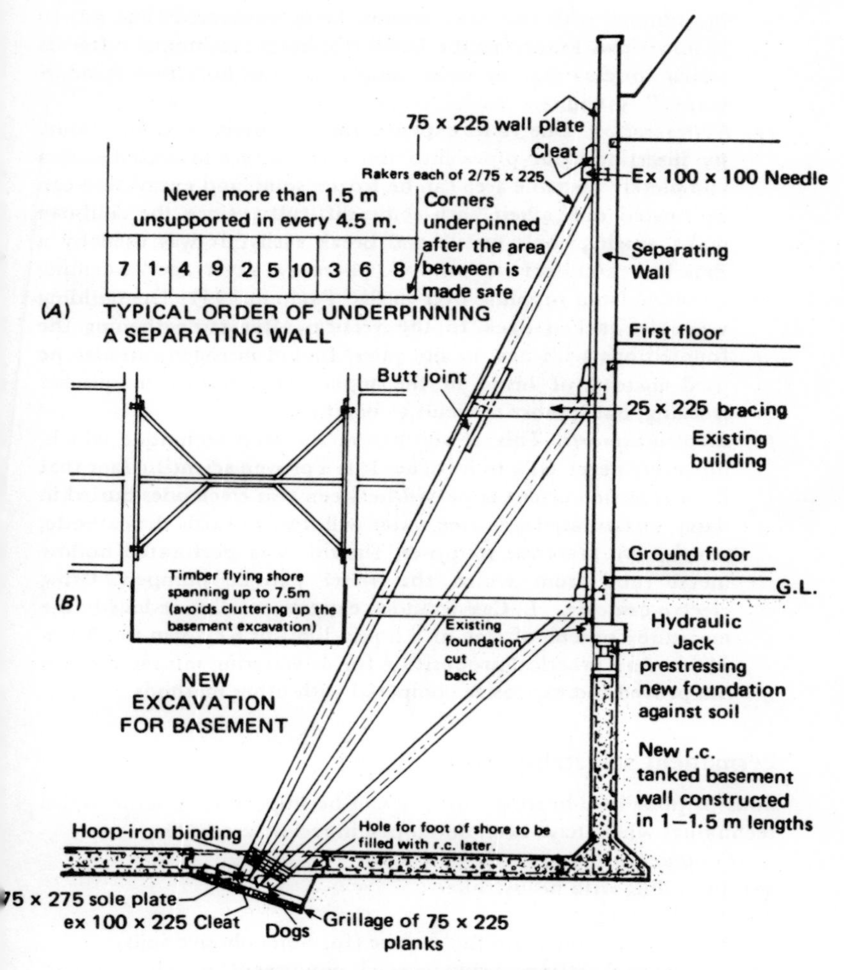

Fig. 102. Shoring and underpinning for a new basement.

(f) *Bentonite slurry.* This is a strange, but effective and now much used technique, by which a trench, as excavated is filled with thixotropic Bentonite (Fuller's earth) slurry. This is quite capable of supporting the sides of a deep trench. Excavation by crane and grab can continue through it and later, concrete poured into the bottom of the trench through a tremie tube causes upwards displacement of the Bentonite, which can then be pumped into the next section being excavated (Fig. 83). It is sometimes known as the ICOS diaphragm technique (after its Italian originators) or more simply as "the hole that is never empty" system.

(g) *Refrigeration.* Due to its expense, this is a rarely used technique. By inserting brine pipes into the surrounding saturated soil, a completely unstable area can be frozen solid, and excavation can be carried out, albeit with some difficulty, since the cold can make cutting tools brittle and break easily. It was used by a Brazilian architect to stop a 24-storey skyscraper weighing 25,000 t from toppling over in Sao Paulo in 1941. The building was later jacked back to the vertical after strengthening the foundations with new in situ piles. Liquid nitrogen can also be used instead of brine in the pipes. This results in quicker freezing, but is more difficult to handle.

(h) *Electro-osmosis.* This is another rarely used technique which, however, never fails to intrigue. It is a proven scientific fact that if an electric current is passed between two electrodes buried in damp or saturated soil, free water will flow towards the cathode, which, for practical purposes, should be a perforated hollow metal tube from which the water can be pumped. Using electro-osmosis, L. Cassagrande experimentally reduced the moisture content of clay at a liquid limit of 67 down to 25% in 100 h. As a practical proposition for de-watering saturated soil it has proved too expensive compared with other methods.

Permanent soil stabilization

Soils of poor load-bearing capacity can be improved by some useful techniques which have been developed in recent years. These are:

1. Compaction

(a) by rolling with heavy rollers;
(b) by ramming;
(c) by vibrating rollers on the surface (for non-cohesive soils);
(d) by vibro-flotation, using special equipment which can consolidate granular soils to 80-90% relative density. (This is based on the consolidating effect of groups of piles on sand as they are driven in, and this simpler method can also be used);
(e) by detonation of explosives (rarely used).

2. Confining—Sheet piling around the site prevents movement of the soil outwards, or in the case of excavations, inwards.

3. Void-filling

(a) by grouting in with cement and sand, or cement and pulverized fuel ash, preferably by pressure grouting using a compressor to force the mixture into the voids (unsuitable for sands and cohesive soils);

(b) by pressure-grouting, using asphaltic or bituminous emulsions which not only fill the voids, but also block the pores, rendering even fine sands impermeable (down to 1 mm particle size);

(c) by pressure-grouting with Bentonite and clay slurry, which is thixotropic.

4. Chemical change

(a) Soil-cement technique. Soil is broken down by mechanically pulverizing it and mixing with dry cement. Water is added to produce a form of weak concrete. (This can be done in situ, but will obviously only strengthen the top layer.) Cement content in dry mix on a weight basis is usually—

6-10% for sandy soils with less than 35% clay;
8-12% for silty soils with less than 35% clay;
10-14% for clays.

(b) Sodium silicate solution is injected into the soil to react with a strong salt solution to form silica gel (Joosten process). For non-cohesive soils ultimate strength is increased considerably.

5. Electrochemical hardening of soils is possible by a process in which a direct electric current is passed between aluminium electrodes inserted into the ground, usually clay. The effect is to reduce moisture content, improve cohesion, decrease the Liquid Limit and prevent swelling, especially frost heave.

6. De-watering. This generally consolidates soils and helps to make them more load-bearing. Various methods are discussed later in this chapter. Wet sand can be consolidated either by de-watering or by saturating it, but the former method is more useful for building purposes. The entry of water can be reduced by Method 2 outlined above and the ground can be made impermeable by Methods 3(b) and 4(b).

B.S. 1924 gives methods of testing stabilized soils, but is mainly concerned with those stabilized by cement, lime or bitumen.

Excavation on sloping sites for buildings

When designing buildings on sloping sites, it is important to consider the implications of plan shape and floor levels for excavation and foundations. Figure 103 illustrates some alternative ways to design a small building utilizing the slope and excavation.

The amount of excavation which is economic depends on the type of soil. In hard rock Fig. 103a might be the best solution and Fig. 103f would be cheaper than Fig. 103e. Figure 103d is

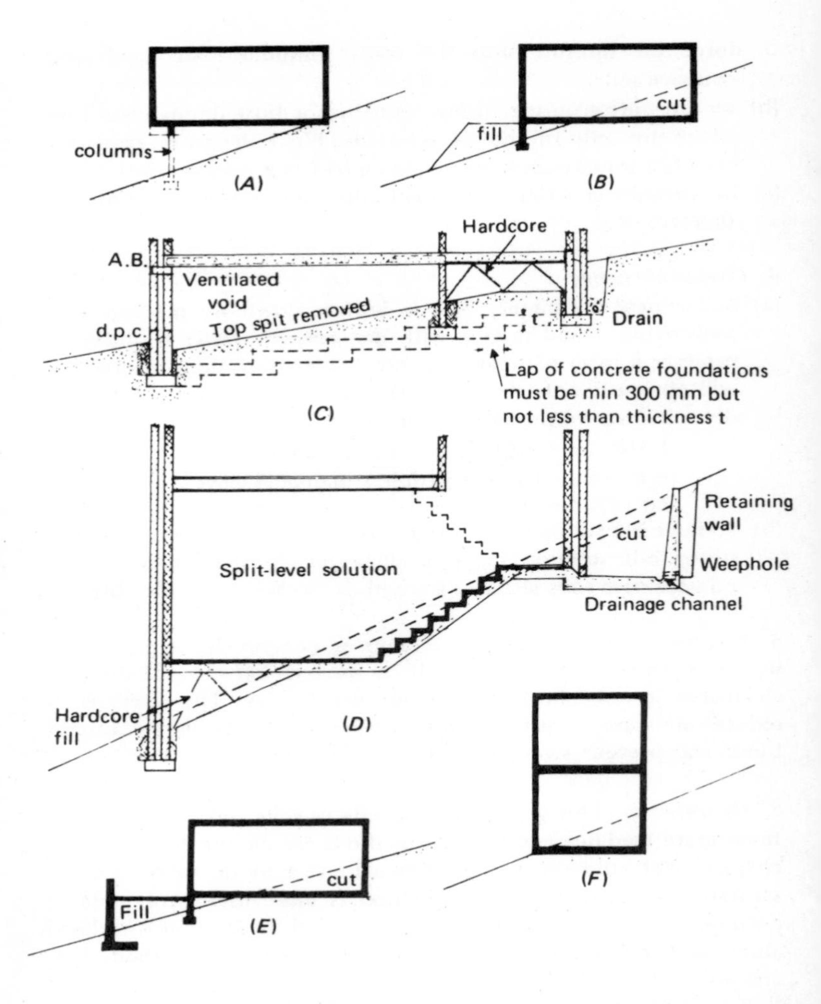

Fig. 103. Alternative ways of dealing with sloping sites.

economical in excavation, a split-level house being a more sensible solution than the builders' commonly used method of building walls up from the sloping ground until a ground floor is achieved level with the highest ground, an expensive and wasteful solution, as in Fig. 103c. This diagram also shows the method of taking excavation for walls and foundations down a slope by cutting horizontal steps. Remember that rock can rarely be cut into sharp-edged steps. Steps should preferably not exceed 300 mm in height, and concrete in the higher foundation must lap the lower foundation for a distance not less than the thickness of the foundation and in no case less than 300 mm (Building Regulations 1972 D7(e)).

Wherever possible it is better to balance cut and fill work as in Fig. 103b and e, and the fill must be very well consolidated if settlement is to be avoided in such situations as Fig. 103e; it is better if the foundation is carried down to undisturbed soil.

The drainage provided across the slope above the building in Fig. 103c and d should be noted. The damp-proofing of parts of a building below ground is more expensive than the use of horizontal D.P.C.s in walls above ground, as shown in Fig. 103c.

Depth to which excavations should be carried

For strip and pad foundations the depth of foundations will depend on:

1. The depth necessary to secure sufficient bearing capacity from the soil and to prevent the soil from squeezing out from under and upwards.
2. The depth necessary to get below the soil subject to high seasonal movement–shrinking in summer and swelling due to moisture and freezing of moisture in winter. In soils like hard chalk and gravel which are little affected, the minimum depth could be 450 mm to the bottom of trenches. In clay, however, it is advisable to go down at least 1 m.

The proximity of drains or other services will often demand a greater depth for foundations than the minimum 450 mm suggested above.

Basement excavations may need to be taken down to great depth so that the excavated material will help balance the weight of a tall building on a fairly low-bearing capacity soil. This is known as the buoyancy principle, and by this means, tower buildings such as the Hilton Hotel and the Shell Building, London, exert very little more pressure on the ground than did the soil excavated from their basements (see Chapter 6).

Raft foundations usually require little excavation. They are often used on land liable to movement and are then designed to be flexible and move with the ground. In these cases, service connections must be kept flexible, or re-connected regularly after movement takes place. Examples of houses built this way can be found on the unstable clay at Cranmore, Isle of Wight. Sometimes the raft is given deeper edge beams which rest on more stable ground and are able to restrict movement of a seasonal nature.

Fig. 104. Block of flats in Italy split in two by foundation failure due to ground heave.

Inspection of excavations for foundations

Notice has to be given to the local Building or District Surveyor when building work is to commence and also when excavations are complete ready for pouring concrete. Depth of excavation and the condition of the soil at the bottom of the trenches etc. must be approved, and the Building Inspector or District Surveyor has the power to order deeper excavation if he thinks it necessary.

Control of water on the site

Water can be a great nuisance on a construction site, turning it into a sea of mud, making excavation difficult, weakening and spoiling the surface of concrete before it has set, and generally making work very unpleasant, more dangerous, and unhealthy for the building operative. The proper control of water is therefore an important part of building work, both at and below ground level.

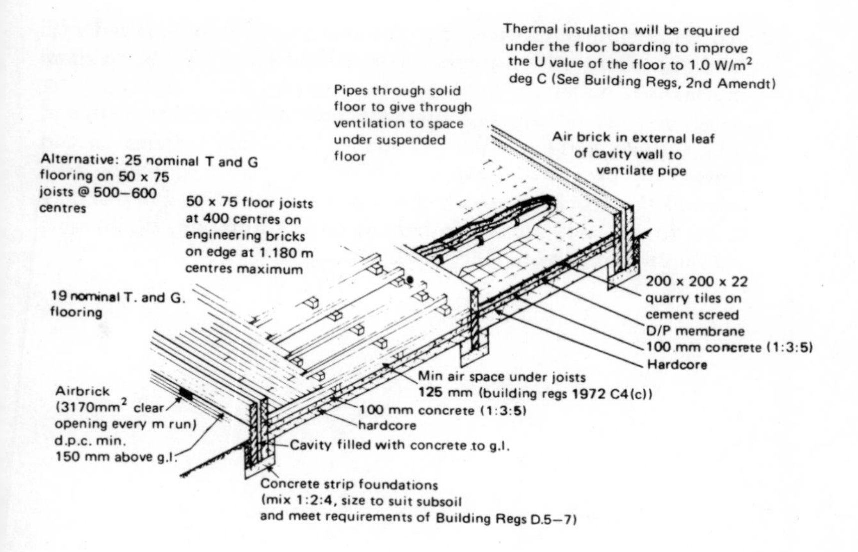

Fig. 105. Suspended timber floors: construction and ventilation to meet requirements of Building Regulations 1972.

Excess of water can be due to:

(a) Rain, or melting snow or hail.
(b) Surface land drainage.
(c) Underground springs or streams, and artesian-type up-flow of water under pressure.
(d) A high ground water level resulting in saturated non-cohesive soils.
(e) Tidal infiltration from sea or tidal estuaries.

Some of the methods used to control the influx of water are as follows:

1. Temporary shelters can be erected over the whole or part of the site by using polyethylene sheets stretched over tubular steel scaffolding. By avoiding delays due to bad weather, this may well pay off handsomely, and such shelters certainly provide better working conditions.

2. To avoid mud and softening of the ground, it can be sprayed with a cheap plastic which forms a surface skin tough enough to stand normal site traffic for the life of the contract.
3. Digging "curtain" drains or ditches round the site can intercept and drain away surface water. Permanent "curtain" drains should always be provided above a building erected on a sloping site. Should the ground be badly drained, a system of land drains may have to be laid, usually in herring-bone formation, discharging into a ditch or stream (Fig. 106).

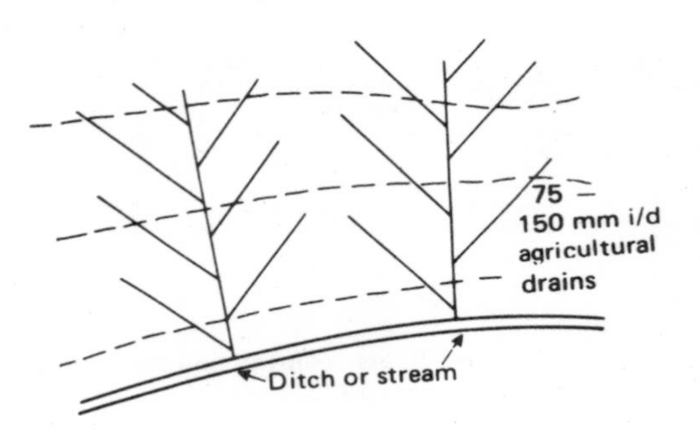

Fig. 106. Herring-bone layout of land drainage.

4. Springs and underground streams encountered in excavation must be piped to a discharge point at a safe distance, or into a storm water sewer. Failing this, pumps may have to be installed, powerful enough to handle the maximum seasonal flow, and to discharge it into a nearby stream or river.
5. Ground water level can be lowered by the use of wells or boreholes with submersible pumps capable of raising 450,000 l/h 30 m vertically. Usually, for trenched foundations, portable pumps (maximum 7.5 m lift) are used to keep the trench bottom dry for concreting. For basement excavations, water can be drained to sumps cut in the soil and then pumped out by pumps which can be of very high capacity if so required. However a more efficient system is to surround the perimeter of the excavation with "well-points" (Fig. 107 and Fig. 86). These consist of fluted steel tubes surrounded by a gauze screen and fitted with a shoe one end and a screwed connection at the top.

TOP SOIL
ORIGINAL WATER LEVEL
RUNNING SAND
SAND AND GRAVEL
GRAVEL
SAND AND GRAVEL
RUNNING SAND
SAND AND GRAVEL

A Wellpoint

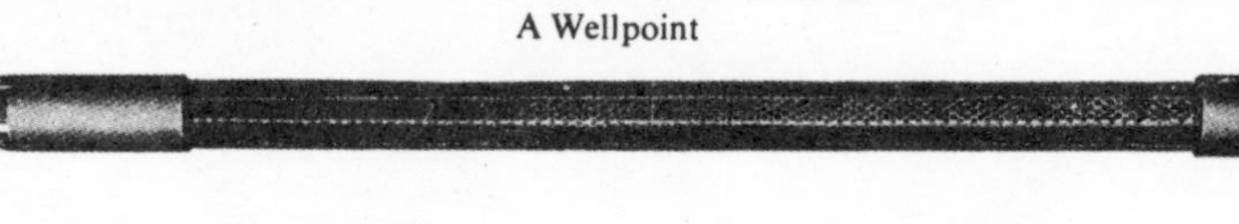

Fig. 107. De-watering an excavation, using "well-points", ring suction main and pump.

These can be sunk into the soil by water jetting down the centre pipe. They are then connected up to a ring suction main to suck ground water through the gauze down to the shoe and up through the central pipe. Spacing of "well-points" is at 750 mm centres or multiples of this, to suit standard connection centres on the 150 mm dia. header pipe. This method of site de-watering is normally carried out down to a maximum depth of 6 m, but by using two or more stages, each with its own pump of high capacity, basement excavations can be kept dry down to 18 m depth. It is important to realize that when ground water level is lowered by pumping, the reduced level is not actually horizontal, but is a series of draw-down curves dipping towards pumping points, the curve varying between shallow with highly permeable soils and steep with soils of lower permeability.

Many of the soil stabilization methods are useful to reduce or prevent the inflow of water. Sheet piling is excellent for this purpose, providing it is driven down into low-permeable soil below the saturated strata, thus providing a tank, or what is known as a *coffer-dam*.

Compressed air has been used to force water out of saturated soil thus allowing it to consolidate, but its use is usually confined to the interior of caissons. These are hollow box or cylinder constructions sunk into saturated soil or water and maintained dry internally by air pressure higher than the pressure of water outside. Caissons are not often seen on building sites, being used mostly for marine and river structures. Workers enter through air-locks, and must spend considerable time being slowly acclimatized to normal air pressure when they leave to avoid attacks of the "bends". Men working in caissons command high wages and the operation is expensive. Depths of 30 m can be reached by this means if required (see also Chapter 7).

6. Tidal infiltration can be dealt with in any of the above ways, but obviously some form of temporary or permanent retaining wall is required to prevent flooding at high tide. It must not be forgotten that a rising head of water can exert considerable upward pressure. The example of Unilever House, Millbank, London, will serve to illustrate this. In the early stages of construction, a tanked basement was built alongside the Thames embankment, but it was found that tidal pressures were being exerted by infiltration underneath, and the whole basement was tending to float like a concrete ship. Only when sufficient building materials were piled up in the basement to counter-balance the upward force of water was stability assured. A basement can also be pinned down against upward pressure by friction piles.

Further references on earthworks, shoring and underpinning, and control of water on site.

B.S. Code of Practice 2003—Earthworks — B.S.I.

Civil Engineering Code of Practice No. 4 — Foundations — Institution of Civil Engineers

Chapter 11—Materials for foundations

Except in those few cases where a level bed of rock is sufficient foundation, concrete is the principal material used for foundations, with some steel for reinforcing in all but the simplest jobs. As the stability of a building depends largely on its foundations, it is important to ensure that these materials are of satisfactory quality for the job they have to do.

Concrete

This is a complex material, and many books have been written on this one subject. Much study is required to understand how to choose the constituents, mix them in the right proportions and ensure that the concrete sets and is cured properly to gain its maximum strength. Some references for detailed study are given later, but the following brief summary explains the essentials:

Concrete is made by mixing:

Cement This is usually Portland Cement to B.S. 12, sold in paper bags containing 50 kg each, or supplied in bulk for large contracts. Other types of cement are used occasionally such as:

(a) *Blast furnace cement* (where readily obtainable), to B.S. 146 and
(b) *Sulphate-resisting cement* to B.S. 4027 where there is a risk of sulphates affecting concrete (see earlier references to this danger).
(c) *Rapid-hardening cement* to enable foundations to take load quickly.
(d) *High-alumina cement* for either or both of the last two reasons or where high temperatures might affect normal concrete.
(e) *Low-heat cement* for very thick foundations where the heat generated as the concrete sets could affect the strength if ordinary cement were used.

Any cement can be made quick-setting by the addition of suitable chemicals, usually proprietary additives. The time needed for set can be reduced from the usual half-hour to almost instantaneous setting if necessary as may be required in underwater conditions.

Water This must be clean and pure, and added in the minimum quantity necessary to ensure hydration of the cement, and to enable mixing and placing to take place. The strength of concrete decreases as the percentage of water increases over this minimum.

Aggregate (to B.S. 882) This can be classified under three headings:

1. *Fine aggregate* Below 5 mm when screened through sieves to B.S. 410. It can be natural sand, crushed gravel or crushed stone, but should be well graded and clean.
2. *Coarse aggregate* 5 mm and above. This can be gravel, crushed stone, or other available inorganic hard waste material.
3. *All-in aggregate* A natural aggregate containing a satisfactory mixture of fine and coarse particles.

Maximum size of aggregate depends on the thickness of concrete being poured, reinforcement spacing and cover, and machine mixing capability. For thinner sections and reinforced concrete foundations, 19 mm down is usually specified, but in mass foundations of great thickness, larger stones known as "plums" are often used to economize in cement.

Concrete mixes

These can vary from strong mixes of 1 part of cement: 1 part of fine aggregate: 2 parts of coarse aggregate by volume (1 : 1 : 2 mix) for high quality reinforced concrete, down to 1 part of cement : 12 parts of all-in aggregate by volume for mass concrete (1 : 12 mix).
Under the *Building Regulations 1972,* the concrete for strip foundations must not be weaker than the mix 1 : 3 : 6, or 1 : 9 if all-in aggregate is used, but few architects would specify any mix weaker than 1 : 3 : 5 or 1 : 8 all-in aggregate for domestic type strip foundations, which allows some margin for safety should a little mud or extra water find its ways into the concrete when it is placed. The most usual mix specified for reinforced concrete is 1 : 2 : 4, although this should be described as C.P. 114-21 N/mm^2, the strength to be attained by the concrete at 28 days.

Concrete is often delivered to the site ready-mixed in specially designed lorries with revolving drum mixers. This often speeds up large volume concrete pouring, and is particularly useful on congested sites.
Reinforcement steel B.S. 4449 specifies rolled steel bars and hard drawn steel wire for concrete reinforcement. Some designers specify cold twisted steel bars to B.S. 4461, and some prefer welded steel mesh to B.S. 4483. For pre-stressed or post-tensioned concrete, steel wire to B.S. 2691 or B.S. 3617 can be specified.

Sizes of reinforcing bars rolled as standard are as follows:

Sizes in mm	Area in mm^2	Weight in kg/m
6	28	.222
8	50	.395
10	79	.617
12	113	.888
16	201	1.58
20	314	2.47
25	491	3.86
32	804	6.31
40	1260	9.87
50	1964	15.41

All are available in 12 m lengths.
For bending dimensions, see B.S. 1478.

Further references on materials for foundations

An Introduction to Concrete— Notes for students — Cement and Concrete Association

Report on Concrete Practice — " " " "
Part 1: Materials and workmanship
Part 2: Site supervision and testing

B.S. Code of Practice C.P. 114 — B.S.I.
Reinforced Concrete in Buildings

Appendix

Examples of the design of simple foundations

1. Design of a strip foundation for a wall

Type A—using unreinforced concrete (Fig. 40).

Required Strip foundation for a 300 mm thick brick wall on soil, with an allowable bearing pressure of 150 kN/m^2, at the depth required for the base of the foundation, and assuming that the total load from wall and foundation acting vertically downwards will be 15 tonnes per metre length.

Method Consider how failure could take place:

(a) By overloading the soil.
(b) By tilting caused by uneven loading and overstressing the soil under one edge. (This can be avoided by ensuring that the resultant of the load is applied centrally on the foundation.)
(c) By diagonal shear at the face of the wall.
(d) By bending forces causing tension, and shear in the concrete when the tensional strength of the concrete is exceeded.
(e) Trouble could occur through movement of the soil under through changes in moisture content. (This would be avoided by taking the foundation down below the depth of soil influenced by seasonal changes.)

So (a) (c) and (d) have still to be dealt with.

Consider a 1 m length of wall and foundation.

Divide total vertical load by allowable bearing pressure of soil q_a.

Total vertical load of 15 tonnes per m = a vertical force P of approximately 150 kN

$$\frac{P}{q_a} = \frac{150}{150} = 1 \text{ m}^2$$

so the strip foundation should be 1 m wide.

By drawing a section of wall and width of foundation to scale and drawing vertical lines at each end downward to meet a 45° line from the face of the base of the wall, a depth of foundation is achieved which will be thick enough to prevent shear, and to prevent bending taking place. This also meets Building Regulations requirements, providing the concrete is not weaker than mix 1 : 9.

Type B—Using reinforced concrete (Fig. 41)

Required Strip foundation for a 300 mm-thick wall on soil with an allowable bearing pressure of 55 kN/m^2 at the depth required for the base of the foundation, and assuming that the total load from wall and superimposed loads, but excluding foundation will be 6.75 tonnes per metre length.

Method Consider how failure could take place:

(a) By overloading the soil.
(b) By tilting due to eccentric loading and, consequent uneven settlement of the soil.
(c) By bending and cracking on underside of the widely projecting portion of foundation which may be necessary on such weak soil.
(d) By reinforcing bars slipping inside the concrete due to inadequate grip length, i.e. permissible average bond stress of concrete with steel, as given in C.P. 114, Table 5, is exceeded.
(e) By excessive bond stress at the point of maximum shear force where Table 5 of C.P. 114 permits a local bond stress up to 1½ times the average bond stress permitted.
(f) By diagonal shear from the foot of the wall downwards through the concrete at about 45°.

Dealing with these one at a time:

1. Taking a 1 m length of wall, find width of foundation necessary to distribute the load over the soil without overloading it:

$$P = \text{approx. } 67.5 \text{ kN } (1 \text{ t} = \text{approx. } 10 \text{ kN})$$

$$\frac{\text{Load}}{\text{Allowable bearing capacity of soil}} = \frac{P}{q_a} = \frac{67.5}{55} = 1.23 \text{ m}^2$$

But the extra weight of foundation concrete over soil removed must be considered, so assume a possible thickness of concrete of 300 mm and estimate weight of base and soil displaced taking weight of reinforced concrete at 2400 kg/m^3, and weight of soil at 2000 kg/m^3.

Weight of foundation less soil
= approximately 1.3 m × 1 m × 0.3 m × 400 kg/m^3
= 156 kg which exerts a pressure of approximately 1.5 kN on the ground.

So, recalculate size of base required for estimated total pressure:

$$P = 69 \text{ kN}$$

$$\frac{P}{q_a} = \frac{69}{55} = 1.25 \text{ m}^2,$$

say, 1.3 m width × 1 m length being considered.

2. Assume resultant of total load is applied central on the foundation for this example.

3. Consider bending of the slab, and calculate maximum bending moment M at face of wall as required by C.P. 114 (see Fig. 41).

Then:

$$M = q_a \times L\left(\frac{B-b}{2}\right) \times \frac{B-b}{4}$$
$$= 55 \times 0.5 \times 0.25$$
$$= 6.875 \text{ kNm}$$

(Actually, full q_a is not developed, but this is a safe figure to work on.)

It is now necessary to decide what quality of concrete mix is to be used, and this will control the permissible stresses to be used in the following calculations.

Assume the following from C.P. 114, Table 5:

Nominal mix	Permissible concrete stresses				
	Compression			Bond	
	Direct	Due to bending	Shear	Average	Local
	kN/m²	kN/m²	kN/m²	kN/m²	kN/m²
1 : 2 : 4	5240	6895	690	827	1240

Permissible tensile stress in steel reinforcement. (From C.P. 114, Table 11) P_{st} = 140,000 kN/m² for bar sizes up to and including 40 mm dia.

Let effective depth

d = assumed total depth 300 mm less concrete cover of 63 mm and less radius of reinforcing bar;

= say 230 mm

Formula for area of steel for the stresses assumed above is:

$$A_s = \frac{M}{0.86d \times P_{st}} = \frac{6.875}{0.86 \times 0.23 \times 140{,}000} = 0.00025 \text{ m}^2 = 250 \text{ mm}^2$$

Say 4 10 mm bars at 250 mm centres per m length of wall foundation. Main bars are rarely spaced at more than 300 mm centres. Some longitudinal reinforcing bars will be required to hold the main bars in place and help distribute the load.

Four 8 mm bars at 400 mm centres will do, wired to the main reinforcement. (Spacing of bars must not exceed $3d$ for main bars and $5d$ for longitudinal distribution bars (C.P. 114) where d = Effective depth.)

Alternatively, a standard welded mesh reinforcement could have been selected (as supplied by B.R.C. Engineering Co. Ltd., or similar). Many engineers select deformed or twisted bars (as supplied by Square Grip Reinforcement Co. Ltd., G.K.N. Ltd., or others similar) some of which are of high tensile steel. The stresses used with these bars will be different (see C.P. 114 and the manufacturer's literature).

Had reinforcement not been used, a mass concrete foundation would have had to be 0.5 m deep to meet the shear requirements.

4. Check average bond stress, i.e. if the grip-length of the bars in the concrete is sufficient to prevent slipping.

For stresses permissible for the steel and concrete specified above:

$$\text{Grip-length} = \frac{P_{st} \times \text{bar diameter}}{4 \times \text{average bond stress}} \qquad \text{(See C.P. 114, Para. 310)}$$

$$= \frac{140{,}000 \times 0.010}{4 \times 827}$$

$= 0.42$ m, i.e. 42 × dia. of bar for these stresses.

With the bars running 1.2 m across the 1.3 m width base and using L-type hooks at each end, there is obviously no problem here.

5. Check local bond stress at the face of the wall where shear is high, as required by C.P. 114, Para. 340d. Para 310b gives the following formula for checking:

$$\text{Local bond stress} = \frac{Q}{l_a O}.$$

Shear force Q at the section will equal total upward force to left of wall face:

$$\frac{P}{B \times L} \times \frac{L\,(B-b)}{2} = \frac{P\,(B-b)}{2\text{B}} = \frac{69\,(1.3-0.3)}{2.6} = 26.5 \text{ kN}$$

O = sum of the perimeters of the bars in a 1 m length of foundation
$= 4 \times 3.14 \times 0.010$ m
$= 0.126$ m

l_a = lever arm of r.c. section = $0.86d$ for the stresses in use.

$$\text{So local bond stress} = \frac{26.5}{0.86 \times 0.23 \times 0.126}$$

$= 1064 \text{ kN/m}^2$

This is below the maximum local bond stress permissible (see above table of stresses).

6. Lastly, check for diagonal shear.

Shear force Q at section of base beyond diagonal shear line will equal total upward soil reaction on that section.

$$Q = \frac{L(B-b-2d)}{2} \times \frac{P}{B \times L} = \frac{P(B-b-2d)}{2B} = \frac{69(1.3-0.3-0.46)}{2 \times 1.3} = 14.3 \text{ kN}$$

Shear stress $q = \dfrac{Q}{L \times l_a}$ (from C.P. 114, Para. 316)

$$= \frac{14.3}{l \times 0.86 \times 0.23}$$

$$= 72 \text{ kN/m}^2$$

This is very much below the permissible maximum shear stress for concrete of 690 kN/m^2. In fact, shear stress is nearly always low in r.c. strip foundations.

2. *Design of a pad foundation for a column in reinforced concrete* (Fig. 42)

Required Foundation for a 300 × 300 mm r.c. column carrying a 70 t load on soil with an allowable bearing capacity of 200 kN/m^2.

Method List possible reasons for failure:

(a) Overloading the soil.
(b) Tilting due to eccentric loading.
(c) Bending and cracking on underside of the widely projecting portion of foundation necessary to spread the load over the soil.
(d) Reinforcing bars slipping inside the concrete due to inadequate grip length.
(e) Excessive bond stress at the point of maximum shear force.
(f) Diagonal shear from the foot of the wall downwards through the concrete at about 45°.

Dealing with these one at a time:

1. Load of 70 t = approximately 700 kN pressure.

$$\frac{P}{q_a} = \frac{700}{200} = 3.5 \text{ m}^2\text{, say } 2 \text{ m} \times 2 \text{ m}$$

Load did not include weight of foundation, so estimate weight of foundation over weight of soil displaced, assuming concrete weighing 2400 kg/m^3 and soil weighing 2000 kg/m^3.

Extra load due to foundation = 2 × 2 × 0.35 × 400 = 560 kg = 5.6 kN approximately.

Recalculate

$$\frac{P}{q_a} = \frac{705.6}{200} = 3.5 \text{ m}^2,$$

therefore 2 × 2 m square base will still be adequate.

2. Assume for this example that the base is centrally loaded so that no moments due to eccentric loading will occur.

3. Consider bending of the slab in one direction, and calculate maximum bending moment M at face of the column (C.P. 114, Para. 340a).
   The soil is not fully stressed up to $q_a = 200 \text{ kN/m}^2$.

Upward pressure from soil will be: $\dfrac{P}{B^2}$

$$\text{So } M = \frac{P}{B^2} \times B\left(\frac{B-b}{2}\right) \times \frac{B-b}{4} = \frac{P(B-b)^2}{8B} = \frac{705.6 \times 1.7^2}{16}$$

$$= 127 \text{ kNm}$$

Assume a possible total depth of r.c. slab of 350 mm. Then effective depth to reinforcement top layer d = 350 mm—cover of 62 mm—1½ diameters of reinforcing bars, say 38 mm = 250 mm. Select concrete mix and steel for reinforcing as in previous example.

$$\text{Formula for area of steel } A_s = \frac{M}{0.86d \times P_{st}}$$

For upper layers of bars in one direction

$$A_s = \frac{127}{0.86 \times 0.25 \times 140{,}000} = 0.00422 \text{ m}^2 = 4220 \text{ mm}^2$$

say fourteen 20 mm bars @ 150 mm centres

For the lower layer of bars in the other direction

$$A_s = \frac{127}{0.86 \times 0.275 \times 140{,}000} = 0.00383 \text{ m}^2 = 3830 \text{ mm}^2$$

say thirteen 20 mm bars @ 160 mm centres.

4. Check if there is adequate grip-length for reinforcing bars:

$$\text{Grip-length required} = \frac{P_{st} \times \text{bar diameter}}{4 \times \text{average bond stress}}$$

$$= \frac{140{,}000 \times 0.020}{4 \times 827}$$

= 0.84 m (42 × dia. of bar, for these stresses).

There will be sufficient grip-length, and L-type hooks can be used.

5. Check local bond stress at face of wall and where shear is high (as required by C.P. 114, Para. 340d).

Local bond stress = $\frac{Q}{l_a o}$ (C.P. 114, Para. 310b)

Shear force Q at the section will equal total upward force to left of wall face:

$$\frac{P}{B^2} \times B \left(\frac{B-b}{2}\right) = \frac{P(B-p)}{2B}$$

$$= \frac{705.6 \times 1.7}{4}$$

$$= 300 \text{ kN}$$

o = Sum of perimeter of top layers of bars in cross-section of slab
= $14 \times 3.14 \times 0.020 = 0.879$ m

$l_a = 0.86d$

$$\frac{Q}{l_a o} = \frac{300}{0.86 \times 0.25 \times 0.879} = 1600 \text{ kN/m}^2$$

This is above the maximum local bond stress permitted, 1240 kN/m^2, so recalculate with more or larger bars.

Say twenty 20 mm bars at 100 mm centres for the top layer and eighteen 20 mm bars at 110 mm centres for the lower layer.
(Note: d is greater for lower layer of bars, and this reduces the local bond stress for this layer to 1123 kN/m^2 when 18 bars are used. For top layer, increasing the number of bars to 20 reduces local bond stress to 1111 kN/m^2.)
Note: The use of high tensile streel bars with greater grip characteristics such as "Square-grip" reinforcement, would reduce the quantity of bars, and therefore the labour of fixing.

6. Check for diagonal shear.

Shear force Q at section of base beyond the diagonal shear line will equal total upward soil reaction on that section.

$$Q = \frac{P}{B^2} \times B \left(\frac{B-b-2d}{2}\right) = \frac{P(B-b-2d)}{2B}$$

$$= \frac{705.6 \times 1.16}{4}$$

$$= 204.6 \text{ kN}$$

Shear stress $q = \frac{Q}{B \times l_a}$

$$= \frac{204.6}{2 \times 0.86 \times 0.25}$$

$$= 476 \text{ kN/m}^2$$

This is below the permissible maximum shear stress of 690 kN/m^2.
The base is therefore proved to be of satisfactory design.

Footnote 1: A useful check as to the probable thickness of a pad base, is obtained from the old formula for punching shear, now no longer included in C.P. 114 but still commonly used. It is:

$$\text{Thickness of base slab} = \frac{P}{\text{perimeter of column} \times 2.5 \times \text{maximum shear stress}}$$

In the example above, this gives:

$$\frac{705.6}{1.2 \times 2.5 \times 690} = 340 \text{ mm thickness}$$

Footnote 2: The reason for the thick cover of 62 mm allowed to reinforcing bars at the bottom of a foundation slab is that it is common practice to lay an initial layer of 25 mm concrete at the bottom of an excavation for a foundation to take up any inequalities, mud or water, before the main cover concrete is poured. The normal cover required under the code of practice is 38 mm for work against earth faces.

References for calculation and design of foundations

The design of structural members, parts 1 and 2 — H. T. Jackson, Architectural Press

C.P. 114. The structural use of reinforced concrete in buildings — B.S.I.

Reinforced Concrete Detailer's Manual — B. Boughton, Crosby Lockwood-Staples

Design and construction of foundations — G. Manning, Concrete Publications Ltd.

Building Regulations 1972 — H.M.S.O.

London Building Acts 1930–39, Constructional Byelaws with explanatory memoranda — G.L.C.

Building Standards (Scotland) Regulations 1971 — H.M.S.O.

The design and construction of engineering foundations — F. Henry

C.P. 110, Pt 1. The structural use of concrete – Design, materials and workmanship — B.S.I.

Reinforced concrete design to C.P. 110 simply explained — A. H. Allen, Cement and Concrete Association

Bibliography

B.S. Code of Practice 2004: Foundations	B.S.I.
B.S. Code of Practice, C.P. 101 Foundations and Substructures	B.S.I.
B.S. Code of Practice 2003, Earthworks	B.S.I.
B.S. Code of Practice, C.P. 3, Chapter V, Loading	B.S.I.
Foundation Design and Construction, M. J. Tomlinson	Pitman
Fundamental Foundations, W. F. Cassie	Elsevier
Foundations of Structures, C. W. Dunham	McGraw-Hill, N.Y.
Design and Construction of Foundations, G. P. Manning	Concrete Publications Ltd.
The Building Regulations 1972	H.M.S.O.
Foundations and Soil Properties, R. Hammond	Macdonald
Specification (latest annual edition)	Architectural Press
B.S. 648, Schedule of Weights of Building Materials	B.S.I.
B.S. Code of Practice, C.P. 114. The Structural Use of Reinforced Concrete in Buildings	B.S.I.

Index

Accuracy in setting out 129
Acidity test 47
Adjustable footing Fig. 92
Aggregates for concrete 146
Analysis of loading 102
Anchored sheet pile retaining walls 102, Figs 52, 65
Area of steel formula 151
Artesian springs 139
Auger Figs 16, 56, 60

Backfill to retaining walls 114, Fig. 85
Bacteria, sulphate-reducing 47
Barrel vaults, foundations to 128
Basements 69, Figs 49, 66, 84–6
Bearing capacity 37
Bench mark 130
Bending moments in foundations Figs 41, 42
Bentonite, use of 133, Figs 64, 83
Boreholes and logs 22, Figs 17, 18
Building Regulations 5, 56
Building (Safety, health and welfare) Regs. 1948 131
Building Inspector 138
Bulb end piles 83, Fig. 55
Bulb of pressure Figs 12, 23, 24
Bulking of sand 13
Buoyancy principle 69, 137
Buttresses Figs 66, 75, 86

Casagrande 5, 134, Table 3
Caissons 93, Fig. 64
Cantilevered piling as retaining wall Fig. 78
Cavities underground 8, 93, Fig. 3
Cement, types available and specification 146
Change in level 135, Fig. 103
Chemical change for soil stabilization 135
Chemical reaction tests on soils 46
Classification of soils 5, Figs 1, 2, Tables 1–3
CLASP foundation Fig. 88
Clay types 15, 39
Coffer dam 142
Cohesion 15, Fig. 11
Cohesive soils 15
Column foundations Figs 42–6, 85
Compaction of soil 91, 93, 134
Compaction tests 41
Compression tests 27, 33, 41, Figs 20–3, 27–9
Concrete mixes 117, 146
Concrete, ready-mixed 69, 146, Fig. 83
  waterproof 114, 117
Cone penetrometer test 24, Figs 14, 16
Conferences on soil mechanics, international xiii
Confined sand 13
Consolidation test 41, Fig. 26

Consolidated undrained triaxial test 42
Constant rate of penetration test 89
Consultant engineers 21
Control of water on site 138
Cores 22, 32
Core extruder 32
Corrosion of steel 47
Cost of soil investigation 21
Coulomb xiii
Counterfort 102, Fig. 75
Cracking in buildings 15, Figs 7, 32, 33, 35–7
Creep of soil Fig. 32
Crib walls 100, Fig. 71
Curtain drains 140, Fig. 103
Cut and fill Fig. 103

Damp-proofing basements 106, 114, Figs 84–6
Datum point and level 130
Deep basements 69, Figs 38, 86
Deflection of structural frames 58, Figs 36, 37
Depth of foundations 9, 15, 19, 47, 69, 77, 78, 83, 137, Fig. 57
De-watering 13, 135, 140, 152, Fig. 107
Diamond drilling rock 22
Dilatency test for cohesive soils 39
District Surveyor 4, 56, 138
Domes, foundations to 128
Drainage, land 140, Figs 106, 107
Drainage of retaining walls Fig. 68
Drainage of basements 112, 114, Figs 85, 86
Drained triaxial test 42
Draw-down of site water by pumping 142, Fig. 86
Dry density/moisture content relationship 40
Drying out of soils 15, 40, Fig. 7
Dumpling Fig. 86

Earthquake forces 100
Eccentric loading Figs 70, 74
Echo-sounding tests 37
Edge-beams Figs 48, 57
Egg-crate grid raft foundation 67, Figs 48, 49
Electrical conductivity tests on soils 37
Electro-osmosis 134
Electrochemical hardening of soils 135
End-bearing piles 19, 72, Figs 55, 56
Erosion 58
Excavation 131, Figs 86, 101–3
Expansion of clay 15, Figs 7, 47
Explosives, use of 131

Factor of safety 33, 37, 88
Failure of foundations 134, 149, 153, Figs 54, 65
Floor heave in cold stores 15
Foundations, buoyancy 69, Fig. 86
  cantilevered Fig. 95
  combined Figs 93, 94
  pad 63, Fig. 42
  piled 69
  raft 67, Fig. 47
  strip 61, Figs 39–41
  under tension 71, 125, Figs 98, 99
Freezing of soil for consolidation 134
Frost, effect of 15, Fig. 8
Fungal attack on timber 73

Geological survey 4
Geology of site 4, 99
Geophysical tests 37
Graphical methods for assessing stability of retaining walls Fig. 70
Gravel 12
Ground water samples 33
Grillage, steel 1
Grip length for reinforcing bars 152

Heel for retaining walls Fig. 72
Herring-bone land drainage 140, Fig. 106
Historical notes xiii

IBIS foundation Fig. 92
Ice lenses, formation of 16, Fig. 8
ICOS diaphragms 134
Impermeable soils 135, Fig. 5
Inaccuracy in setting out 129
Inclined filter drains 114
Insect attack on timber 73
Inspection of foundations 138
Interlocking sheet piles 73, Fig. 52

Jacks, hydraulic 59, 81, 89, 134, Figs 50, 59
Jacking-in piles Fig. 50
Jacking-up in cases of settlement 59, 134
Joints, columns to foundations Figs 43–6
Joints, in basement walls and floors 106, Fig. 86
Joosten process 135

Kentledge 33, Fig. 62

Laboratory tests on soil 32, 39
Land drainage 140, Figs 106, 107
Limestone 5, 9, 11
Liquid limit 39, Fig. 25
Load assessment 56
Loads, dead, live, wind, lateral 56, 97
Load factor 33, 37, 88
Loam 19
Local bond stress 152
London clay 15, 39
Longitudinal reinforcement, strip foundations 63

Machinery foundations, anti-vibration and sound insulating 125, Fig. 97
Made-up ground 19
Magnetic variations test 37
Masonry retaining walls 100, 117, Figs 67–9
Materials for foundations 145
Mechanical plant 120, 131
Mine workings, records of 4
Mining settlement Fig. 33
Mohr's circles 42, Fig. 29
Moisture content 39
Moving loads 56
Mud 19

National Coal Board records 4
Newmark influence charts 47
Non-cohesive soils 12

Oedometer 41, Fig. 26
Ordnance Survey 4, 130
Optical measuring equipment 131
Organic matter test 46

Pad foundations 65, Figs 12, 42
calculations for 153
Particle size distribution 40, Fig. 2
Partition wall foundations 56, 63, Fig. 34
Passive resistance of soil Fig. 65

Peat 18
Penetrometer 24, Fig. 16
Penetration 24, 89
Permanent shuttering 132, Fig. 86
Permeability 13, Fig. 5
pH value 47
Pile boring 83, Fig. 60
 clusters and caps 23, Figs 13, 63
 drivers 81, Figs 58, 59
 foundations 69
 testing 83, Figs 61, 62
 types and materials 73, Figs 50–9
Piling problems 91
 tenders and economics 93
Plastic flow of clay 16, 39
Plastic limit 39, Fig. 25
Plasticity index 39
Plate loading test 33, Figs 21, 22
Pneumatic structures, foundations for 127
Polings Fig. 86
Post-tensioning Figs 77, 96
Pre-fabricated buildings, foundations for 119, Figs 87–92
Pressure grouting 117, 134, Fig. 79
Principles of foundations 53
Puddled clay 114
Pulverised fuel ash (P.F.A.) 134
Pumps, well-point 140, Figs 86, 107
 sump 114, 140

Quicksands 13, Fig. 6
Quick-setting cement additives 145

Raft foundations 67, 137, Figs 47, 48
Reactor foundation Fig. 77
Reamed-out bases for piles (belled) 83, Fig. 56
Refrigeration of soil 134
Reinforcement types 146
Retaining walls 97, Figs 65–86
Revetments 100, 103, Figs 80, 81
River bank walls 100
Rock 8
Rock-bolts, rock-pinning, rock-stitching 11, Fig. 4
Rock testing 22, 37
Rotational shear 16, Figs 10, 11, 65

Sand 12
 quick 13, Fig. 6
Sandstones 8, 11
Saturated soils 15
Scour, effects of 103, Fig. 81
Sea walls 100, 103, Figs 80, 81
Seismic refraction test 37
Selection of foundation type 61
Setting out 129, Fig. 100
Settlement, 16, 56, Fig. 9
Shear 42, 53, Fig. 32
Shear box test 43, Figs 30, 31
Sheet piles 78, 132, Figs 52, 54
Shelby sampler 22
Shores, flying and raking Figs 86, 102
Short bored piles 78, Fig. 57
Shrinkage, linear, test 40
Sieves, B.S. 410 40
Silt 17, 39, Fig. 1
Site tests 21, 27, Figs 14–16, 19–21, 61–2
Site investigation report 47, Figs 17, 18, 29, 31
Sizes of reinforcing bars 147
Sleeper walls and supports for timber floors 63, Figs 90, 105
Sloping sites 135, Figs 82, 103

Slope failures 16, Figs 11, 82
Soil-cement technique 135
Soil investigation 21
Soil mechanics 1
Soil profiles Fig. 17
Soil samples 32
Soil stabilization 134
Soldiers Fig. 86
Sound insulating foundation 125, Fig. 97
Split tube sampler (split spoon) 22, Figs 15, 16
Stabilized soils, tests for (B.S. 1924) 135
Steel for reinforcement 146, 151
Stepped foundations 136, Fig. 103
Stone 8, 11
Strengthening existing foundations 124, Fig. 96
    in concrete 151
Stresses, permissible 151
    in reinforcing steel 151
Strip foundations 61, Figs 12, 39–41
    calculations for 149, 150
Sulphates, tests for 47
Sulphate-bearing clay 18, 48
Survey, site 3
Swallow holes 8, Fig. 3

Tanked basements 69, Fig. 49
Tell-tales 4
Tension structures, foundations to 71, 125, Figs 98, 99
Tests, laboratory 39
    site 21, 83, 88, 89, 135, Figs 61, 62
Terzaghi 1, 134
Thixotropic suspensions (Bentonite) Figs 64, 83
Tidal infiltration 142
Timber piles 73, Fig. 51
Timbering of trenches 17, 132
Tolerances 130
Top spit 19, 129
Tree roots 129, Fig. 7
Trenches for foundations 132, Figs 39, 101
Trial pits 22
Triangular diagram for soil classification 5, Fig. 1
Triaxial tests 41, Figs 27–9

Unconfined compression test 27, Fig. 20
Underpinning Fig. 102
Undrained triaxial test 42

Vane test 27, Fig. 19
Vibration insulated foundation 125, Fig. 97
Vibro-flotation 134

Walings Figs 86, 101
Wash-boring 22, 141, Fig. 16
Water pressure on basements Fig. 84
Waterproofing basements 114, Figs 85, 86
Water table 23, 35
Weather, effect on foundation building 138
Weight of building materials 56, 150
Welded mesh reinforcement 146, 151
Well-points 140, Figs 86, 107
Wind load effect 56, Fig. 32